30-SECOND
BIOLOGY

30-SECOND
BIOLOGY

The 50 most thought-provoking
theories of life, each explained
in half a minute

Editors
Nick Battey
Mark Fellowes

Contributors
Nick Battey
Brian Clegg
Phil Dash
Mark Fellowes
Henry Gee
Jonathan Gibbins
Tim Richardson
Tiffany Taylor
Philip J. White

Illustrator
Steve Rawlings

IVY PRESS

First published in the UK in 2016 by
Ivy Press
210 High Street, Lewes,
East Sussex BN7 2NS, UK
www.ivypress.co.uk

British Library Cataloguing-in-
Publication Data
A CIP catalogue record for this
book is available from the
British Library.

ISBN: 978-1-78240-389-0

This book was conceived,
designed and produced by
Ivy Press
Publisher **Susan Kelly**
Creative Director **Michael Whitehead**
Editorial Director **Tom Kitch**
Commissioning Editor **Stephanie Evans**
Project Editor **Joanna Bentley**
Designer **Ginny Zeal**
Illustrator **Steve Rawlings**
Glossaries Text **Charles Phillips**

Typeset in Section

Printed and bound in China

10 9 8 7 6 5 4 3 2 1

CONTENTS

INTRODUCTION
Nick Battey & Mark Fellowes

It's a little over 200 years since the word

'biology' was invented to identify the part of the natural world capable of reproducing and maintaining itself. In the beginning it attracted vitalist thinking (the assumption that life required some mysterious 'force' to sustain it), as people sought to account for the unusual abilities of living organisms. But the development of cell theory, physiology and the theory of evolution in the nineteenth century gradually helped to fill the gaps about how life worked. In the twentieth century genetics, biochemistry, molecular biology and developmental biology established clear mechanistic knowledge of the way life runs itself, showing how the features and processes characteristic of organisms are regulated during the life of the individual and passed on to offspring. At the same time, ecology, evolutionary biology and biogeography became key disciplines for understanding the behaviour and interactions of populations of organisms with each other and their environment. The subdisciplines of zoology, botany, microbiology and virology elucidated the details of how life within the different kingdoms worked, while taxonomists and systematists explored its hierarchies and origins.

Twenty-first century science

The domains of biology continue to expand, making it no exaggeration to claim that it is *the* science of this century; there is little of global importance that is not touched by it. Biomedicine, including regenerative medicine and medical genetics, is exploiting biological knowledge to give ever-increasing control over life, death, illness and disease. At the same time, the other aspects of biology allow us to engage with the crucial problems that threaten to overwhelm society – climate change, population growth, pollution, food shortages, erosion of natural resources and species invasions. In a sense the integrative, ecological dimensions to biology take on the consequences of advances in human

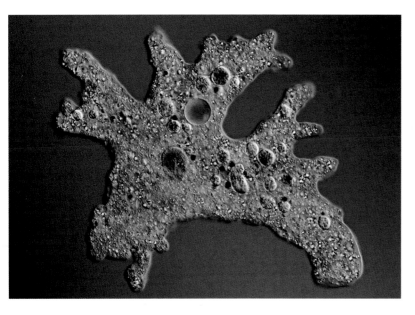

Amoeba proteus is a tiny one-celled organism often found in freshwater environments.

biology. As the dominant discipline of the Anthropocene (the proposed new geological epoch that reflects the current human domination of our planet, see page 150), biology conditions what we can do and shapes how we deal with the results of our actions. Almost everything that matters is affected by biology.

The one notable absence from this brief appraisal is culture. Broadly speaking, culture can be considered to be the consequence of human consciousness, passed on through the generations not by genes but by one form or other of mental transmission. Psychology deals with the human mind, and we have not considered it part of biology in this book because it is a distinct subject with largely separate traditions to biology; and biology itself is anyway big enough as originally defined. But an obvious prediction is that rigorous biological explanations of the way the mind works, and eventually of the way culture works, will be definitive advances when viewed from the perspective of 2050: that the *30-Second Biology* of that date will include the science of the mind as a dominant feature. A different guess might be that biological explanations of the mind and the culture it generates are prosaic: what matters is not what causes culture but what it means to us, a realm in which biological explanations may be of little interest. Not so, though, the central place

The impact of climate change and human settlements on the alpine habitat of snow leopards makes them an endangered species.

of biology in the future life of the human species. How we deal with a rapidly growing (and ageing) population, the destruction of habitats, and how we regulate what we do with our biological power, will be critical issues that it will take all our biological and cultural skills to address.

How this book works

In *30-Second Biology* each topic is clearly and concisely presented on one page in a punchy single paragraph: the **30-second theory**. For an even quicker overview, there is the **3-second dissection** alongside – the key facts encompassed in a single sentence. Then the **3-minute dissection** fleshes this out, addressing a thought-provokingly quirky or intriguing aspect of the subject. Each chapter also contains the biography of a pioneer or high achiever in the field – people such as Norman Borlaug, the developer of high-yielding, semi-dwarf wheat, who is credited with saving millions from starvation and hailed as the 'father of the green revolution'.

The book begins with **Life**, a scan of the major groups of living organisms. It then discusses **Genes**, the blueprint of life, and, in **Genes to Organisms**, looks at how the information encoded by genes is transformed into the cells and tissues of living organisms. Next **Growth & Reproduction** are considered, focusing on these processes in plants, animals and bacteria. **Energy & Nutrition** looks at how energy is converted into life, and how living processes maintain and are maintained within bodies. Finally, the chapters on **Evolution** and **Ecology** discuss how life arose, how organisms live together and the peculiar strains placed on these relationships by the phenomenal growth of the human species. It is this dominance that threatens to destabilize everything on the planet; the growth of biology is one reason for it, but understanding how life works and how organisms interact seems to us something intrinsically beautiful. Let's see if that category still exists in 2050.

LIFE ◑

LIFE
GLOSSARY

biofilm Self-sufficient community of bacteria in which different species may cooperate, some recycling the wastes of others. Dental plaque is a biofilm.

cell Smallest unit of an organism, typically but not always consisting of a nucleus and cytoplasm encircled by a membrane. Many microscopic organisms – such as bacteria and yeasts – consist of just one cell.

centrioles Organelles (compartments within a cell) found in pairs close to the nucleus in animal cells. Centrioles play a key role in cell division.

chloroplast A plastid (type of organelle) found in the cells of green plants, in which photosynthesis occurs.

choanoflagellates Free-living single-celled microscopic eukaryotes believed to have been the evolutionary ancestors of animals. They are flagellate – they contain whip-like organelles known as *flagella*.

cloning Asexual reproduction of genetically identical copies of an original organism or cell. Cloning occurs naturally: plants' and animals' body cells are ultimately clones of an original fertilized egg. Cells are also cloned in the laboratory:

for example, the nucleus can be removed from an egg and replaced with the nucleus of a cell of the type to be cloned.

cyanobacteria Single-celled prokaryotic organisms that derive energy through photosynthesis. Also known as blue-green bacteria, they are the earliest known lifeform on Earth – their fossil record in western Australia dates to 3.5 billion years ago.

cytoplasm The part of a cell that surrounds the nucleus and is enclosed by the cell's outer membrane – the cellular membrane.

DNA Deoxyribonucleic acid, a molecule that carries the coded genetic information that transmits inherited traits. DNA is found in the cells of all prokaryotes and eukaryotes.

Dolly the sheep The first mammal to be cloned from an adult cell. In 1996 a team from the Roslin Institute and biotech firm PPL Therapeutics cloned Dolly, a domestic sheep, from a sheep's mammary gland cell using the nuclear transfer method (adding new genetic material into a cell from which the original genetic material had been taken out). Dolly was genetically identical to the sheep whose genetic material was added to the mammary gland cell. She lived more than six and half years from 5 July 1996 to 14 February 2003.

eukaryote Organism or cell that has a discrete nucleus.

gene Unit of heredity, located on a chromosome. Genes consist of DNA, except in some viruses where they are made of RNA.

genetically modified An organism that has been modified genetically, often to produce desirable traits – for example, resistance to pests in a plant.

genome Complete genetic material in an organism or a cell.

mitochondrion Organelle found in most cells in which energy production and respiration take place. Plural: mitochondria.

nucleus Central organelle within most eukaryotic cells that contains genetic material. It is enclosed by a double membrane – the nuclear membrane.

organelle Compartment or structure within a cell.

photosynthesis Process by which green plants produce their food (sugars and starches) from water and carbon dioxide, giving off oxygen as a byproduct.

The process is powered by light energy from the sun, which plants harness through the chlorophyll found in the chloroplasts within their cells.

prokaryote Single-celled organism that does not have a discrete nucleus or contain any other organelles.

protists Group of distantly related mainly microscopic organisms, each usually consisting of a single cell. Some, such as algae, contain chloroplasts and are more like plants, while others, such as amoebae, are more like animals. A third subgroup, which includes yeasts, are closer to fungi.

RNA Ribonucleic acid, a molecule – found in all living cells – that plays a key role in the synthesis of proteins. In some viruses, RNA rather than DNA functions as the carrier of genetic information.

spore A single-celled reproductive unit found in some plants and fungi.

symbiogenesis The theory that eukaryotic organisms evolved through associations with prokaryotic bacteria.

ORIGINS OF LIFE – VIRUSES

the 30-second theory

3-SECOND DISSECTION
Many attempts have
been made to define 'life'.
Perhaps it's like jazz –
you can't define it, but
you know it when you
experience it.

3-MINUTE SYNTHESIS
The discovery of
mimiviruses suggests
that viruses, simple as
they are, evolved from
more complex forms of life.
What pre-viral life was like
is completely mysterious,
as the simplest bacteria are
more complex than even
the largest known viruses,
and organized differently.
The largest known virus,
Pandoravirus, discovered
in 2013, is very different
from other viruses and
might represent a hitherto
unknown form of life.
Fortunately, *Pandoravirus*
only infects amoebae.

Life appeared on Earth more than 3.5 billion years ago, less than 1 billion years after the Earth itself formed. Nobody knows what the first lifeforms – simple, self-reproducing chemical systems – were like. However, complex organic compounds are known to occur on the surfaces of comets and icy grains in space. Most of the Earth's water is believed to have come from comets, so it is possible that the ingredients for life arrived on Earth from space. The simplest organisms living today are viruses. Some say they hardly count as living at all, because they are completely inert unless they infect living cells, on which they are reliant for reproduction and spread. A virus consists of a small amount of genetic material enclosed in a protein coat. When a virus infects a cell it hijacks the cell's machinery to copy its own genetic material (whether DNA or RNA) and create protein coats. Eventually the infected cells burst, spreading thousands of new viruses to infect new cells. We know viruses best as agents of human disease – from smallpox to Ebola, influenza to HIV – but viruses infect organisms of all kinds, even bacteria. Although viruses are so small they cannot be seen except in an electron microscope, some recently discovered viruses, the mimiviruses, are large enough to be infected by other viruses.

RELATED TOPICS
See also
BACTERIA
page 18

PROTISTS
page 22

GLOBAL PHYLOGENY
page 126

3-SECOND BIOGRAPHIES
LOUIS PASTEUR
1822–95
French scientist who, unable to
find a causative agent for
rabies (now known to be a
virus), speculated correctly
that it was too small to be seen
with an ordinary microscope

DMITRY IOSIFOVICH
IVANOVSKY
1864–1920
Russian botanist, the first to
discover viruses, specifically,
the tobacco mosaic virus

30-SECOND TEXT
Henry Gee

Viruses make infected cells burst, spreading infection to new cells. They cause diseases from the common cold to poliomyelitis.

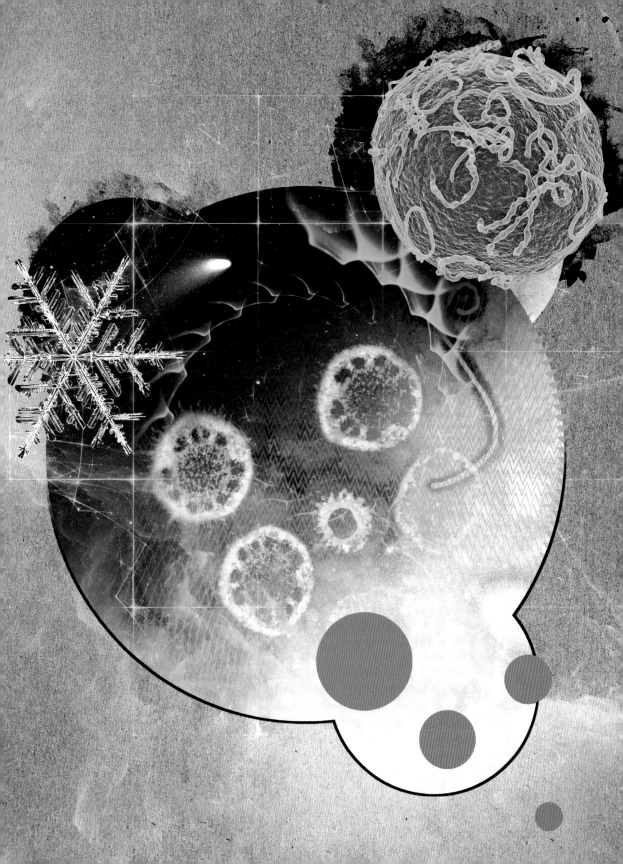

ARCHAEA

the 30-second theory

Archaea are tiny, single-celled creatures superficially similar to bacteria. Like bacteria, they are prokaryotes. That is, their cells contain neither a discrete nucleus nor compartments or 'organelles' such as mitochondria or chloroplasts, as are found in eukaryotes (protists, plants, animals and fungi). Once thought only to be associated with exotic environments such as hot springs, recent research shows them, like bacteria, to be ubiquitous, living in most environments. Archaea produce methane that comes out of the back ends of cows (and people). There are even archaea specific to the human navel. Unlike bacteria, however, no archaeon is known to cause disease in humans. On the whole, archaea have a sense of mystery about them – most cannot be cultured in laboratories, and in many cases we only know of their existence from DNA discovered in the environment and sequenced later. Studies of this DNA show archaea to be fundamentally different from bacteria; in many ways they are closer to eukaryotes, and modern ideas suggest that eukaryotes evolved from within the archaea.

3-SECOND DISSECTION
Carl Woese suggested that archaea, bacteria and eukaryotes formed three 'domains' of life. With archaea closer to eukaryotes, three has become two.

3-MINUTE SYNTHESIS
The very latest research shows that archaea discovered in sediment from the floor of the Arctic Ocean contain genes only otherwise found in eukaryotes. These 'Lokiarchaeota' confirm the view that eukaryotes evolved from archaea. It seems likely that an archaeon formed the basis of the nucleus of the eukaryotic cell, with other components such as mitochondria, chloroplasts and centrioles having evolved from bacteria with which the archaeon became associated.

RELATED TOPICS
See also
BACTERIA
page 18

LYNN MARGULIS
page 20

GLOBAL PHYLOGENY
page 126

3-SECOND BIOGRAPHY
CARL WOESE
1928–2012
American biologist who first used gene sequencing to show that archaea formed a separate 'domain' of life, distinct from bacteria.

30-SECOND TEXT
Henry Gee

Archaea are not limited to places like hot springs, as once thought, but are found in most environments.

BACTERIA

the 30-second theory

Wherever you look you will find bacteria. Although these single-celled creatures are so small that individual examples are only just visible with a light microscope, their combined mass exceeds that of all plants and animals. For every cell in your body there are ten bacteria, mostly in your gut and on your skin. Bacteria thrive in a wide range of environments: from deep in the Earth's crust to the edges of space, in every gram of soil and every drop of water. Although some cause diseases such as tuberculosis, leprosy, meningitis, cholera and bubonic plague, many live in harmony with animals and plants and are vital for cycling nutrients. Bacteria were the first known living things on Earth, appearing in the fossil record more than 3,400 million years ago. Although not much to look at – their range of form is limited to rods, spheres, spirals and a few others – and the insides of their tiny cells show little in the way of substructure compared with animal, plant or protist cells, bacteria are distinguished by their eclectic metabolism. They decompose waste; convert nitrogen from the atmosphere into a form plants and animals can use; produce oxygen, without which we could not breathe; and convert milk into yoghurt and cheese.

3-SECOND DISSECTION
Bacteria were discovered by pioneer microscopist Antonie van Leeuwenhoek in 1676 – but weren't seen again for another century.

3-MINUTE SYNTHESIS
Bacteria can form multicellular 'biofilms' in which cells aggregate to exchange nutrients. Biofilms accumulate on the ocean floor – and in the lungs of patients with cystic fibrosis. Some of the earliest fossils visible to the naked eye are layered, cushion-like structures called stromatolites and made up of cyanobacteria. These still live today in a few parts of the world where the sea is too salty for creatures that might graze on them.

RELATED TOPICS
See also
ARCHAEA
page 16

DEVELOPMENT &
REPRODUCTION: BACTERIA
page 74

MUTUALISMS
page 122

3-SECOND BIOGRAPHIES
ROBERT KOCH
1843–1910
German scientist who identified the bacteria that cause cholera, anthrax and tuberculosis and founded our modern understanding of infectious disease

PAUL EHRLICH
1854–1915
German scientist who invented the first effective antibiotic against a bacterial infection, arsphenamine – a 'magic bullet' against syphilis

30-SECOND TEXT
Henry Gee

Robert Koch (top) and Paul Ehrlich were pioneers in the study of bacteria.

1938
Born Lynn Petra Alexander in Chicago

1957
Graduates from the University of Chicago with a bachelor of arts in Liberal Arts; marries astronomer Carl Sagan

1960
Moves to University of Wisconsin, swiftly followed by moves to Berkeley (where she obtained her PhD in 1965) and then Brandeis in Massachusetts (1964)

1964
Divorced from Sagan

1966
Moves to Boston University

1967
Publishes 'On the Origin of Mitosing Cells', the article that would become a landmark in her theory of symbiogenesis

1967
Marries crystallographer Thomas N. Margulis

1970
Publishes her book *Origin of Eukaryotic Cells*

1978
Symbiogenesis theory proved experimentally

1980
Divorced from Margulis

1988
Appointed distinguished professor at the University of Massachusetts, Amherst

2011
Suffers a stroke and dies five days later

LYNN MARGULIS

Every science needs its revolutionary thinkers, for today's outlandish ideas become tomorrow's textbook orthodoxy. There were few as revolutionary as Lynn Margulis, whose ideas about symbiogenesis, initially dismissed as crazy, are now the cornerstone of modern biological thought. Born to a large Jewish family in Chicago, the fiery and precocious Lynn entered the University of Chicago aged just 16, and her first academic paper, on the genetics of the protist *Euglena*, was published when she was 20. Her notoriety began in 1966 with a paper on the origin of eukaryotic cells, which she suggested had evolved from associations of bacteria. She proposed that the organelles of cells, such as chloroplasts and mitochondria, evolved as separate organisms but became assimilated into a new kind of organism, the eukaryotic cell. It was more than a decade before her ideas were substantiated by significant experimental evidence, and we now know that they are largely correct. It turns out that chloroplasts – small, green bodies in plant cells in which photosynthesis takes place – have their own DNA, revealing that they were once descended from cyanobacteria (once called blue-green algae). Mitochondria, for their part, are small bodies that generate much of the energy required by cells; they also have their own DNA, and are distantly related to bacteria called proteobacteria. Like many people with controversial ideas, Margulis did not stop there. With James Lovelock (born in 1919) she became a vocal proponent of the 'Gaia' hypothesis, according to which the Earth is a single, self-regulating system, and, more controversially, she contended that the human immunodeficiency virus (HIV-1) was not a cause of AIDS (Acquired Immune Deficiency Syndrome). She married and divorced twice, and later reportedly said that it was not humanly possible to be a first-class scientist, wife and mother all at the same time.

Henry Gee

PROTISTS

the 30-second theory

3-SECOND DISSECTION
Many protists such as amoebae, paramecia and algae live freely in ponds and puddles and – unlike bacteria – are easily seen with a simple microscope.

3-MINUTE SYNTHESIS
All protists – with animals, plants and fungi – are eukaryotes. That is, their cells are large, with the genetic material contained in one or more discrete nuclei, separated from the cytoplasm, which may contain various other bodies or organelles such as mitochondria and chloroplasts. These cells are much larger and more complex than those of bacteria and archaea, collectively known as the prokaryotes. Eukaryotes evolved 1-2 billion years ago, probably from some form of archaea.

The protists comprise a varied group of mainly microscopic organisms, united only in that each usually consists of a single cell. Some, such as amoebae and *Paramecium*, are more like animals, whereas others, the algae, contain chloroplasts and are more like plants. Yet others, such as yeasts and slime moulds, are closer to fungi. A few are notorious agents of human disease, such as *Plasmodium*, which causes malaria, and *Trypanosoma*, which causes sleeping sickness; and some are pests, such as the dinoflagellate algae that cause 'red tides' (algal bloom). Being unicellular doesn't mean that protists skimp on complexity. Some algae 'swallowed' other protists in their evolution and have become remarkably complex, with the DNA of up to four ancestral organisms. Others, such as diatoms, radiolarians, coccolithophores and foraminiferans, produce exquisite shells of calcite or silica. Perhaps the most remarkable protists are the rare warnowiid dinoflagellates, which have complex stinging capsules and tiny 'eyes' with the equivalents of a lens and a retina – all within a single cell. Some protists form multicellular associations: seaweeds are multicellular algae, and when slime-moulds are starved they get together to form a mobile, slug-like creature.

RELATED TOPICS
See also
ARCHAEA
page 16

MUTUALISMS
page 122

GLOBAL PHYLOGENY
page 126

3-SECOND BIOGRAPHIES
ANTONIE VAN LEEUWENHOEK
1632–1723
Dutch draper and lensmaker who made the first microscope and was the first to observe protists

LYNN MARGULIS
1938–2011
American biologist who developed the theory of 'symbiogenesis' – according to which many forms of life, including eukaryotic cells, originated by the fusion of other, simpler forms

30-SECOND TEXT
Henry Gee

Despite being single-celled, protists can be complex and sometimes harmful.

FUNGI

the 30-second theory

Many fungi live almost unnoticed in the cracks and crevices of the world. Along with animals and plants, fungi form the third great group of multicellular eukaryotes – but they are usually only noticed by their fruiting bodies, some of which we call mushrooms and toadstools. Most of the time they exist as networks or 'mycelia' of very fine threads, or 'hyphae'. Hyphae spread through the soil or water in which the fungus lives. If they meet hyphae from a fungus of the same species, they may unite to create a sexual phase, which will then produce a fruiting body. When mature the fruiting body sheds spores, which germinate into new hyphae. Like animals, fungi live by breaking down organic matter. They include many pests and diseases such as moulds, the rusts and smuts of crops, athlete's foot, ringworm and Dutch elm disease, as well as a disease caused by chytrid fungus which is threatening to wipe out amphibians worldwide. On the other hand, many green plants could not live without the mycorrhizae attached to their roots, harvesting soil nutrients; fungi produce antibiotics; and without yeasts to ferment plant material, we'd have no wine or beer.

RELATED TOPICS
See also
PLANTS
page 26

MUTUALISMS
page 122

3-SECOND DISSECTION
Fungi are more closely related to animals than plants; lichens are a symbiosis between fungi and algae.

3-MINUTE SYNTHESIS
Fungal hyphae are almost microscopically fine and can spread far and wide. Therefore it shouldn't be a surprise that fungi include some of the largest, heaviest and oldest of all living things. A single individual of *Armillaria bulbosa* in the USA occupies almost 15 hectares (37 acres), weighs more than 10,000 kg (22,000 lb) and is at least 1,500 years old – but because it is made of a network of microscopic, underground hyphae, the totality is almost impossible to grasp.

3-SECOND BIOGRAPHY
ALEXANDER FLEMING
1881–1955
Scottish biologist who accidentally discovered penicillin, an antibiotic produced by a fungal contamination in the cultures of bacteria he was studying

30-SECOND TEXT
Henry Gee

Tasty mushrooms, deadly toadstools, wine and blue cheese ... without fungi the world would be a far duller place.

PLANTS

the 30-second theory

3-SECOND DISSECTION
Plants are living things made of many cells. They produce their own food by using sunlight to convert water and carbon dioxide into sugars.

3-MINUTE SYNTHESIS
Green plants stay in the same place throughout adult life, which makes them easy prey for animals. Plants fight back by armouring their cells with tough, near indigestible walls of cellulose and lignin, and producing a wide range of bitter-tasting poisons – some of which, such as aspirin, are now used as medicines. On the other hand, plants entice animals with rewards such as floral nectar to attract pollinators, aiding reproduction.

Of all living things, plants are probably those we rely on most for our existence, and yet are most likely to take for granted. The majority of the material we use to feed, clothe and house ourselves comes from plants. Oil and plastics are made from the remains of plants that died hundreds of millions of years ago. Green plants are responsible for the oxygen we need to breathe, produced through photosynthesis. In this process green plants use the sunlight-trapping green pigment chlorophyll to combine water and carbon dioxide to create sugars and starches. Although human agriculture relies on just a few plants, mainly grasses such as wheat, rice and millet, there are several hundred thousand species of green plants. The first green plants evolved from simple algae and appeared on land sometime before 400 million years ago. These were simple, thin-stemmed plants with spore sacs on their tops. Plants soon evolved the hard tissues we call 'wood' and by 360 million years ago the first forests spread across the Earth. The first trees were fern-like forms, only later to be replaced by conifers. The flowering plants that dominate the landscape today first appeared in the time of the dinosaurs, between 100 and 200 million years ago.

RELATED TOPICS
See also
ANIMALS
page 32

COEVOLUTION
page 122

3-SECOND BIOGRAPHIES
CARL VON LINNÉ (LINNAEUS)
1707–78
Swedish botanist obsessed with the sex lives of plants who invented the system of classification on which modern systems are based

IRENE MANTON
1904–98
British botanist who worked on ferns and algae and pioneered electron microscopy to study the fine structure of plant cells

30-SECOND TEXT
Henry Gee

From ferns to trees, plants are vital for our lives. Forests have been with us for a staggering 360 million years.

ANIMALS

the 30-second theory

It's easier to recognize an animal than to define it. Or is it? Many animals look like plants; some have bizarre shapes or life habits; some are invisible to the naked eye; and many will be unfamiliar to anyone except a specialist scientist. Many people and most children will equate the term 'animal' with 'mammals' – the group of animals that contains ourselves and most of our domestic animals, such as cats, dogs, sheep, pigs and cows. But mammals are just one part of a larger group of animals, the vertebrates, which also includes birds, fish, amphibians and reptiles. Looking further afield, the closest relatives of vertebrates include – surprisingly – starfish and sea-squirts. Other animals include the arthropods – jointed-legged creatures such as insects, crustaceans and spiders. They also include the annelids, segmented worms such as earthworms and leeches, and molluscs, such as clams and squid. Even simpler animals are the jelly-like sea anemones and jellyfish; and the sponges are the simplest of all. There are around 37 kinds or 'phyla' of animal, some of them extremely obscure. Single-celled organisms such as the amoeba are nowadays classed as protists rather than animals.

RELATED TOPICS
See also
PROTISTS
page 22

FUNGI
page 24

GLOBAL PHYLOGENY
page 126

3-SECOND DISSECTION
Animals are living things, each of many cells, though they reproduce via single-celled eggs and sperm. They feed on other living things.

3-MINUTE SYNTHESIS
Animals evolved from single-celled protists between 650 and 550 million years ago. Nobody knows how many animal species there are, though most – at least 1 million – are insects, which is why, when biologist J.B.S. Haldane was asked what he could deduce about the mind of God, he reportedly quipped 'an inordinate fondness for beetles'. Perhaps, though, the most successful animals in terms of population and biomass are small marine crustaceans such as copepods and krill.

3-SECOND BIOGRAPHIES
GEORGES CUVIER
1769–1832
French zoologist who made one of the first serious attempts at a branching classification of animals

LIBBIE HYMAN
1888–1969
American zoologist whose textbooks on animal anatomy and classification are still standard works today

30-SECOND TEXT
Henry Gee

Mammals occupy just one branch on the animal tree of life, whose roots go back to single-celled protists 600 million years in the past.

SYNTHETIC LIFE

the 30-second theory

The first 'synthetic' organism

was announced in the journal *Science* in 2010. J. Craig Venter and colleagues had synthesized the million-base genome of a mycoplasma (a very simple living organism) from scratch and inserted it into a cell of a different mycoplasma whose DNA had been removed. The new organism, which contained genetic 'watermarks' illustrating its synthetic nature, was capable of dividing in the laboratory. The new organism isn't strictly synthetic, however, because it was based on an evolved template and used an existing cell as a chassis. In this way, the technology represents less something totally new than a fusion of existing technologies such as cloning, laboratory synthesis of DNA, and – as in Dolly the Sheep – inserting new genetic material into a cell whose own genetic material has been removed. Scientists remain very far indeed from creating a new, self-replicating living organism from scratch. Perhaps the key missing ingredient is the living cell in which the genetic material is inserted. Our knowledge of how cells live, use resources, void waste and divide is still too rudimentary to allow that. On the other hand, the capability of creating new virus particles is well within current technology.

3-SECOND DISSECTION
Synthetic biology draws from disciplines as diverse as evolutionary biology and electrical engineering, thinking of life as a 'circuit' of interacting molecular elements.

3-MINUTE SYNTHESIS
Humans have been creating new breeds of animals and plants since the dawn of agriculture. Introducing foreign genetic material into plants and animals to create genetically modified organisms (GMOs) is a shortcut, allowing the production of traits that would take many generations to select. Synthetic life is a further step – the creation of organisms whose genes are designed entirely by computer. A potentially unsettling idea, as their behaviour once realized might be hard to predict or control.

RELATED TOPICS
See also
ORIGINS OF LIFE – VIRUSES
page 14

ANIMALS
page 28

DNA, RNA, PROTEINS
page 36

3-SECOND BIOGRAPHIES
WERNER ARBER
1929–
Swiss biologist who, with Hamilton Smith and Daniel Nathans, shared a Nobel prize in 1978 for the discovery of restriction enzymes

J. CRAIG VENTER
1946–
American biologist and entrepreneur, the first to sequence the human genome as a private venture, and to create something close to a synthetic lifeform

30-SECOND TEXT
Henry Gee

Test tube life? Venter and colleagues inserted a synthetic genome in a biological cell.

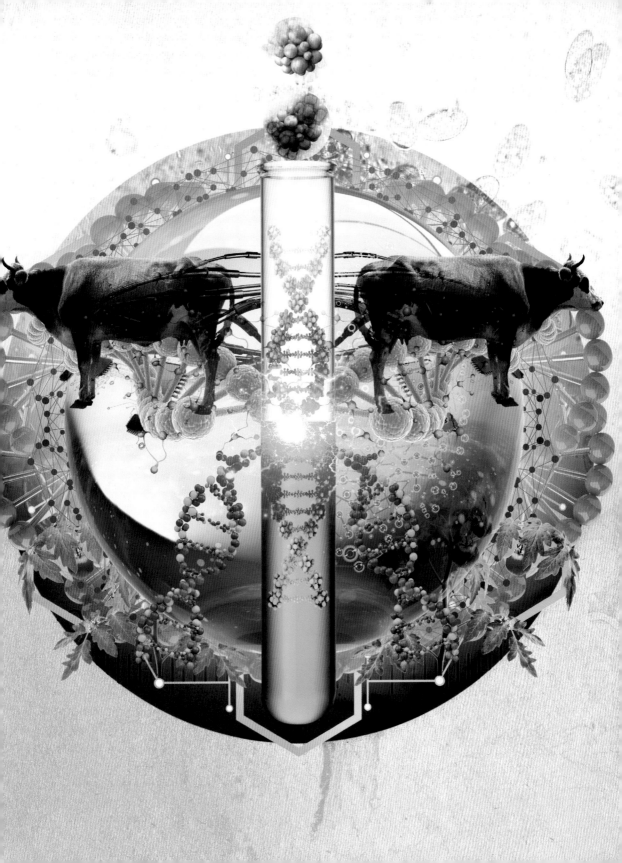

GENES

GENES
GLOSSARY

allelle Also known as an allelomorph, an alternative form of a gene, situated in the same place on a chromosome as the original form of the gene would have been.

amino acids Water-soluble organic compounds that are constituent parts of proteins. Of around 24 amino acids involved in the making of proteins, ten cannot be made by the human body and so must be included in the diet: they are called essential amino acids.

chromosome Thread-like structure that bears the genes that carry genetic information. Chromosomes are found in the nucleus of eukaryotic cells (those that have a discrete nucleus); they consist mainly of DNA, with some RNA and a core of proteins. A prokaryotic cell (one without a discrete nucleus) has a single chromosome made entirely of DNA.

deep ancestry An organism's genetic inheritance from the distant past, perhaps from hundreds of millions of years ago.

DNA Deoxyribonucleic acid, a molecule that carries the coded genetic information that transmits inherited traits. DNA is found in the cells of all prokaryotes and eukaryotes.

eugenics Pseudoscience of 'improving' human populations through genetic modification or selective breeding. Possible means of doing this might include forcibly sterilizing people carrying so-called 'defective' genes. The term was coined by British natural scientist Francis Galton.

gene Unit of heredity, located on a chromosome. Genes consist of DNA, except in some viruses where they are made of RNA.

genetic drift Random fluctuations in the frequency of genetic variation in a population. It is one of the mechanisms by which evolution works.

genome and genotype The genome is the complete group of genes or genetic material in an organism or a cell. Genomics is the study of an organism's genome, focusing on its evolution, function and structure. The genotype is the genetic makeup of an organism.

inclusive fitness Theory in evolutionary biology that explains the origin of altruistic behaviour. It states that individuals with a certain percentage of shared genes cooperate in order to promote the passing on of genes to the next generation. The associated kin selection theory proposes that animals behave altruistically and socially when it benefits relatives' reproductive success.

metabolome Complete set of small molecule chemicals in an organism.

migration Movement of populations (groups). In evolutionary terms migration may move genes from one population to another.

mutation Change in a gene's structure, resulting from changes in bases of DNA or the rearrangement, deletion or addition of sections of genes or chromosomes.

natural selection The process through which those organisms that are best adapted to their environment will survive and have more offspring. Imagine a mixed population of black and white insects in an environment where the black ones could hide from the birds that eat them more easily than the white ones. Because more white insects are eaten, fewer white insects reproduce; meanwhile being black enables black insects to survive and have more progeny – the black insects are better adapted to the environment. Over time the white insects will die out and eventually the population will consist entirely of black insects. In the theory of English naturalist

Charles Darwin, natural selection was one of the key mechanisms – along with genetic drift, migration and mutation – by which evolution worked.

proteins Organic compounds that are essential components of living cells. Their molecules consist of chains of amino acids.

proteome Complete group of proteins expressed by the genome.

RNA Ribonucleic acid, a compound – found in all living cells – that plays a key role in the synthesis of proteins. In some viruses, RNA rather than DNA functions as the carrier of genetic information.

species Group of organisms whose members can interbreed and produce fertile offspring. Species is the eighth category in the scientific classification system, beneath Genus.

X-ray crystallographer Scientist who investigates the structure of biomolecules by interpreting the atomic/molecular structure of their crystal forms.

DNA, RNA, PROTEINS

the 30-second theory

3-SECOND DISSECTION
DNA encodes the blueprint of life; RNA copies and deciphers the code into molecular machines – proteins.

3-MINUTE SYNTHESIS
In humans, around 98 per cent of DNA is non-coding – containing no genes. This was originally considered 'junk DNA', but we now know that many of these sequences are functional. Some encode RNA molecules which are not translated into proteins (for example, tRNA); some encode microRNAs, which regulate transcription. Some are relics of old genes, and some are hitchhikers – selfish and viral genetic elements, sneaking into DNA and replicating with the cell. This non-coding DNA has created an 'Enigma code' for scientists to decipher.

Deoxyribonucleic acid (DNA)

stores the instructions for how to build proteins – the molecular machines of every living cell. In 1953 Watson and Crick worked out the structure of DNA: two sugar-phosphate strands running in opposite directions, with paired bases at the centre, coiled into the famous double-helix structure. This complementary structure provides a copying mechanism that enables cells to replicate. As a relatively stable molecule, DNA makes an excellent storage device, but for the information it encodes to be useful it needs to be transcribed and translated into proteins – which, as enzymes and structural components, are the biologically active elements of cells. Messenger ribonucleic acids (mRNAs) are copies of the coding regions of DNA, and move from the nucleus to ribosomes in the cell cytoplasm. Each ribosome holds an mRNA in place and recruits complementary transfer RNAs (tRNAs) to the site. Attached to each tRNA is an amino acid – the building blocks of proteins. These amino acids are held closely together, forming chemical bonds – while tRNA moves down the mRNA, translating the code, three bases at a time. Once complete, the amino acid string is released and folds into a complex and exact three-dimensional structure, forming a protein.

RELATED TOPICS
See also
EPIGENETICS
Page 42

GENOMICS AND OTHER 'OMICS
Page 44

CELLS AND CELL DIVISION:
page 54

3-SECOND BIOGRAPHY
ROSALIND E. FRANKLIN
1920–58
An often-overlooked English chemist and X-ray crystallographer whose expertise helped reveal the structure of DNA

30-SECOND TEXT
Tiffany Taylor

The celebrated double helix of DNA – entwined sugar-phosphate strands enclosing paired bases – contains the code for making proteins.

MENDELIAN GENETICS

the 30-second theory

3-SECOND DISSECTION
Ever wonder why you're the only one in the family with blue eyes? Mendel gave us the means to understand with his experiments on pea plants.

3-MINUTE SYNTHESIS
The significance of Mendel's work was eventually appreciated in 1900, following its rediscovery, leading to revolutionary advances in genetics. For example, Thomas H. Morgan and his laboratory group at Columbia University were able to demonstrate, based on experimentation with fruit flies, that genes are located on chromosomes and that the gene is the unit of inheritance, giving a physical basis for understanding Mendel's observations. Morgan's research won him the Nobel Prize for Medicine in 1933.

Mendelian genetics are so-called because of pioneering work by Augustinian friar Gregor Mendel. Between 1856 and 1863, he carried out an elegant set of experiments using selective breeding of pea plants over many generations. This explained how traits are passed from one generation to the next. Mendel noticed that offspring of the pea plants would possess a number of their parents' traits in fixed proportions. For example, flower colour: when two plants with white flowers were crossed, the offspring always had white flowers; yet when two plants with purple flowers were crossed, there was variation in the frequency at which offspring possessed white flowers, from zero, to 1 in 4. Mendel found these frequencies were predictable based on the lineage's family history and was able to deduce that such traits were inherited discretely, as dominant or recessive. A dominant trait will always be observable in the organism that carries the gene coding for it. If recessive, individuals can carry the gene, but the effects will be masked by the dominant. If both parents carry one recessive and one dominant gene, the offspring may express a trait that neither parent has – in this example, parents with purple flowers can produce offspring with white flowers. These revolutionary insights went on to form the cornerstone of modern genetics.

RELATED TOPICS
See also
DNA, RNA, PROTEINS
page 36

POPULATION GENETICS
page 40

EPIGENETICS
page 42

3-SECOND BIOGRAPHIES
WILLIAM BATESON
1861–1926
English geneticist who carried out a series of plant-breeding studies that replicated Mendel's results

REGINALD C. PUNNETT
1875–1967
English geneticist best remembered as the creator of the Punnett square: a tool used by biologists that applies Mendelian genetics to predict the expected ratio of offspring given parental genotype

30-SECOND TEXT
Tiffany Taylor

Using pea plants, Gregor Mendel worked out how genetic traits are passed down generations.

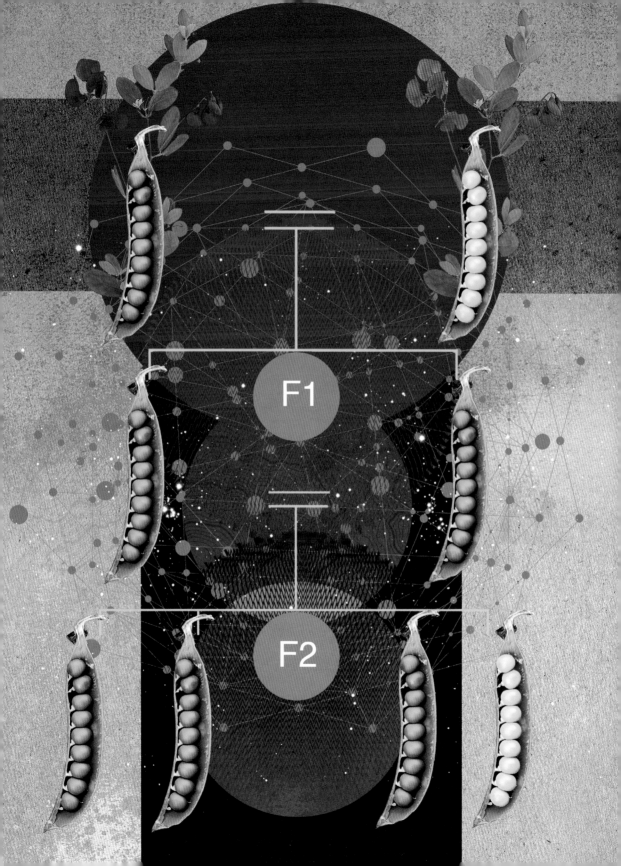

POPULATION GENETICS

the 30-second theory

Species live in groups called populations that can interact and breed with each other. Differences between individuals within a population can be caused by genetic variation that arises through random mutation within genes – these different variants of genes are called alleles. By studying how often different alleles occur within and between populations, scientists can track evolutionary change. Four main processes drive changes in allele frequency. Selection increases the frequency of alleles coding for traits that confer a survival advantage and decreases those that impose a disadvantage. Genetic drift governs random fluctuations in allele frequencies that will not necessarily promote better survival – this process has a larger effect on small populations. Migration causes variation in allele frequencies by introducing new alleles into a population – a process known as gene flow. Mutation (changes to the DNA code through random 'mistakes' during cell copying) creates new allele variants. Population geneticists study the individual influences of and complex interactions between these fundamental principles, addressing many disparate problems from understanding the persistence of relatively common genetic diseases to providing practical conservation strategies.

RELATED TOPIC
See also
MENDELIAN GENETICS
page 38

3-SECOND DISSECTION
Selection, drift, migration and mutation drive changes in allele frequency. Using these fundamental principles we can understand evolutionary change in populations over time.

3-MINUTE SYNTHESIS
The sickle-cell gene causes the disease sickle-cell anaemia, which is comparatively more frequent in parts of Africa. In theory, an allele that has negative effects on fitness should be purged from populations by natural selection – why then, is it maintained? Carriers (those that carry the gene but do not suffer from the disease) are more tolerant to malaria. Therefore in areas of high malaria incidence, natural selection acts to maintain the allele in the population.

3-SECOND BIOGRAPHIES
SEWALL G. WRIGHT
1889–1988
American geneticist who formalized the concept of genetic drift using mathematical theory

RONALD A. FISHER
1890–1962
English evolutionary biologist who pioneered mathematical techniques in population genetics

JOHN B.S. HALDANE
1892–1964
British-born scientist who mathematically formalized the law of natural selection

30-SECOND TEXT
Tiffany Taylor

Population geneticists keep track of evolutionar change by following the trail of variations caused by gene mutations within groups ('populations').

EPIGENETICS

the 30-second theory

Every cell in your body shares identical DNA; and yet a heart cell will perform very distinct functions compared to a kidney cell. This is because signals in the environment in which a cell exists can cause genes to be turned on or off. These differences between genetically identical cells are described as epigenetic. Three systems modify gene expression in this way: DNA methylation, histone modification and RNA-associated silencing. DNA methylation attaches a methyl group to specific points in the DNA, changing the structure of the DNA so that it cannot be transcribed and translated into proteins. Histone modification occurs when an acetyl or methyl group is attached to specific amino acids in the histone proteins around which DNA is wound in the cell nucleus. This changes the structure of the histone and alters access of transcriptional regulators to the DNA, activating or inactivating large sections of it. Finally, RNA can inactivate genes by RNA-associated silencing, which might cause histone modification or DNA methylation or cause DNA to become so tightly wound that it can no longer interact with transcriptional regulators. Such epigenetic changes are part of normal cellular functioning, and their disruption can lead to genetic diseases.

3-SECOND DISSECTION
Spot the difference –
when it comes to gene
expression, there's more
to it than what's written
in your DNA.

3-MINUTE SYNTHESIS
In 1983, links were discovered between cancer and epigenetics. Diseased tissues from patients with colorectal cancer were found to have less methylation than healthy tissues. Methylation acts to switch off genes, therefore low methylation can cause cells to grow indefinitely – a hallmark of cancerous cells. However, excessive methylation has also been shown to disrupt genes that act to suppress tumours, therefore healthy tissues need a balance.

RELATED TOPICS
See also
DNA, RNA, PROTEINS
page 36

CANCER
page 84

3-SECOND BIOGRAPHY
MARY FRANCES LYON
1925–2014
English geneticist who discovered the phenomenon of X-inactivation, an epigenetic process whereby one X-chromosome is always inactivated in females to compensate for the problem of double X-dosage

30-SECOND TEXT
Tiffany Taylor

Epigenetic changes lie behind the way a cell becomes specialized as a brain cell or liver cell, but also behind diseases like cancer.

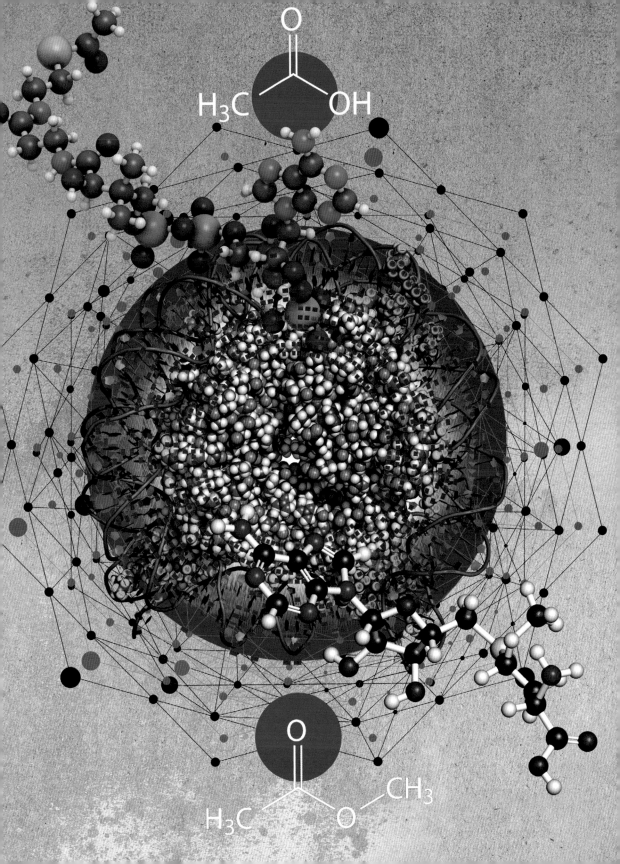

GENOMICS & OTHER 'OMICS

the 30-second theory

A genome is the complete DNA of an organism and provides all the information for how to build and maintain it. Genomics is a field within genetics that aims to understand how genome sequence and structure translate into function. After the advent of sequencing technologies in the late 1970s there was an explosion of data – at the time of writing this entry there were 13,036 complete sequenced genomes, all freely accessible to members of the public. However, the rate at which data were generated far exceeded our ability to understand them, and this misalignment has led to an 'omics revolution. The 'omics refer to disciplines within biology that aim to deliver functional understanding of the discipline to which the suffix is attached – these include genomics, proteomics (the study of the proteome, the entire set of proteins expressed by the genome) and metabolomics (the study of the metabolome, the entire set of small molecule chemicals within an organism). These relatively new yet highly productive fields represent a shift towards understanding the whole organism in terms of the quantitative effects of its genes and gene products. Where previously genes were considered in isolation, now we have the technology to analyze an entire living system at the molecular level.

3-SECOND DISSECTION
The human genome is made up of 3 billion bases, but these are just a jumble, until we can understand what they mean.

3-MINUTE SYNTHESIS
Sequencing DNA has become so cheap and easy that data are being generated more quickly than they can be properly analyzed. In biology, this big data revolution has been quick, leaving little time to adapt. Researchers need more computing power, better organization of data and a more efficient way of moving their data around. The cost of such computational hurdles is threatening to create a bottleneck in biological research.

RELATED TOPIC
See also
DNA, RNA, PROTEINS
page 36

3-SECOND BIOGRAPHIES
FREDERICK SANGER
1918–2013
British biochemist who deciphered the structure of insulin and developed a method for sequencing DNA

WILLIAM JAMES (JIM) KENT
1960–
American scientist who wrote a computer program that enabled the publicly funded Human Genome Project to assemble its fragments before a rival private corporation – ensuring the genome would be freely available to all.

30-SECOND TEXT
Tiffany Taylor

The human genome is a working blueprint for making a human – if you can follow the instructions.

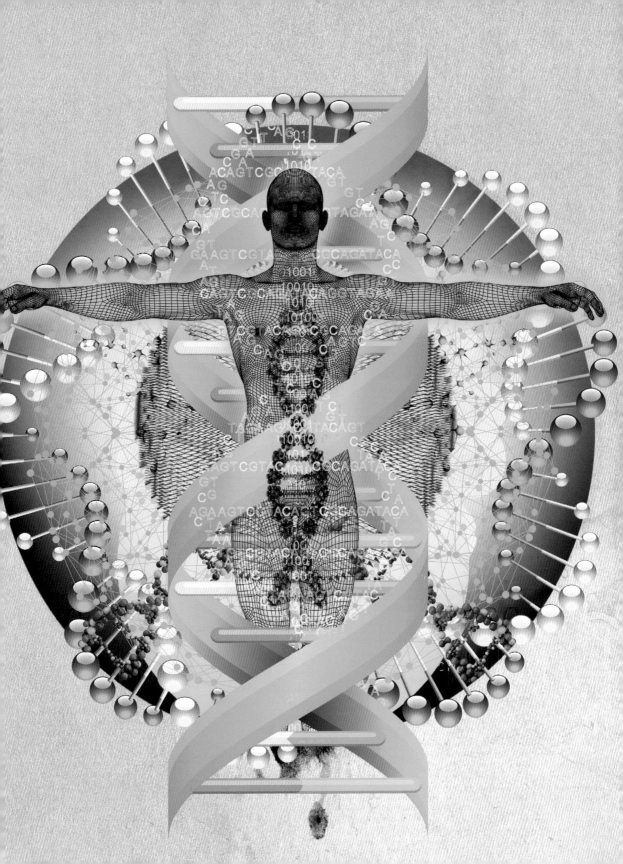

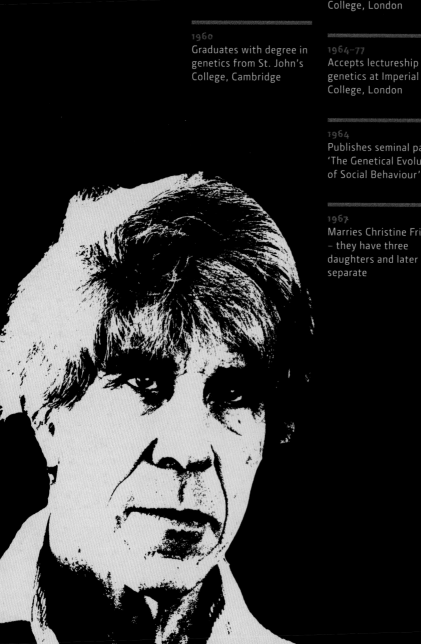

1936
Born in Cairo, Egypt, to Archibald and Bettina Hamilton

1960
Graduates with degree in genetics from St. John's College, Cambridge

1968
Earns a PhD from the London School of Economics and University College, London

1964-77
Accepts lectureship in genetics at Imperial College, London

1964
Publishes seminal paper 'The Genetical Evolution of Social Behaviour'

1967
Marries Christine Friess – they have three daughters and later separate

1978-84
Appointed professor at Michigan University

1980
Elected as Fellow of the Royal Society of London

1984
Returns to England as Royal Society research professor and a fellow of New College, Oxford

1988
Awarded Darwin Medal of the Royal Society of London

1994
Meets partner (Maria) Luisa Bozzi

2000
Dies in Middlesex Hospital

BILL HAMILTON

Bill Hamilton was known as a

mathematician and biological theorist; however, he was also a notable naturalist – and his fascination with this subject was evident from an early age, when he would spend his spare time collecting butterflies and other insects. His parents bought him a copy of E.B. Ford's *Butterflies* from the Collins New Naturalist Series, and it was here that he first learnt about the fundamentals of natural selection, genetics and population genetics that would later shape his career. After reading Ford, he asked for a copy of Darwin's *On the Origin of Species* as a school prize.

However, it was as an undergraduate at the University of Cambridge, after reading Ronald Fisher's *The Genetical Theory of Natural Selection*, that he began to unify Darwin's principles of natural selection with population genetics. Both Fisher and J.B.S. Haldane had recognized that there were many incidences of helping behaviour in nature, but there was a paradox in understanding how such behaviours could evolve, given that natural selection should favour individuals that maximize their own fitness – something Darwin also noted. This intrigued Hamilton and later led to his most influential ideas regarding kin selection and inclusive fitness.

Cooperation is fundamental to life and can be found at all levels of biology – genes cooperate to form genomes, cells cooperate to form organisms and individuals cooperate to form societies. Hamilton developed an elegant theoretical framework, employing a theorem developed by Sewall Wright, to explain the evolution and maintenance of cooperation, which showed how genetic relatedness between interacting individuals can determine the fitness of shared genes rather than individuals; this later became known as Hamilton's rule. His insights were integral to the development of a gene-centred view of evolution, which was later popularized by Richard Dawkins in *The Selfish Gene*.

A scientific risk-taker, Hamilton's radical ideas were often not immediately appreciated, although in time they were widely recognized. He found it difficult to find advisors and funding, and for a while had no desk and worked in public parks and railway stations. This made him question his own sanity, something he alluded to in his first volume of collected papers *Narrow Roads of Gene Land*.

Hamilton later became interested in the origin of HIV and went to the Democratic Republic of Congo to conduct field work. He died soon after his return, aged 63, after contracting malaria.

Tiffany Taylor

GENETIC SCREENING
the 30-second theory

Until very recently, predicting

your medical future was a statistical art. Your age, your physical characteristics, what you ate and drank and whether you smoked were all compared to known populations and your risk of falling victim to disease estimated. But now you can spit in a tube, send it to a commercial company and around six weeks later receive a personalized report that provides an estimate of your likelihood of suffering from, or passing on, diseases. While there is a range of approaches to genotyping, these firms are most likely to search for variants of Single Nucleotide Polymorphisms (SNPs, colloquially known as 'snips'). These are mutations at a single point in the genome, and can be associated with increased probability of disease or how we might respond to drugs, as well as inherited physical traits, psychological conditions and even our deep ancestry. New associations are frequently being discovered, making this an ever more powerful way to understand our futures. Should we know how we might die? Would it make us change our behaviour to reduce risk factors? Or would it encourage pessimism and nihilism, as there is little we can do to change what is predetermined? More importantly for society, should insurers, pension providers or employers have access to this information?

3-SECOND DISSECTION
Personal genetic testing will revolutionize health care; while this is likely to be medically beneficial, the consequences for society are less clear.

3-MINUTE SYNTHESIS
Although there are immediate concerns about the consequences of having access to personal genetic data, how will we allow such information to be used in the future? One can imagine a time where it is used to determine what is 'best', devaluing those who are different. There is a widening gulf between society's position on the morality of eugenics and the possibilities emerging from lightning-fast changes in technology.

RELATED TOPIC
See also
MENDELIAN GENETICS
page 38

3-SECOND BIOGRAPHY
ANNE WOJCICKI
1973–
CEO and co-founder of 23andMe, a highly influential personal genomics firm that has already genotyped more than 1 million people

30-SECOND TEXT
Mark Fellowes

Would you be happier if you knew the likelihood – based on genetic screening – of your developing or passing on diseases? It might change how you live.

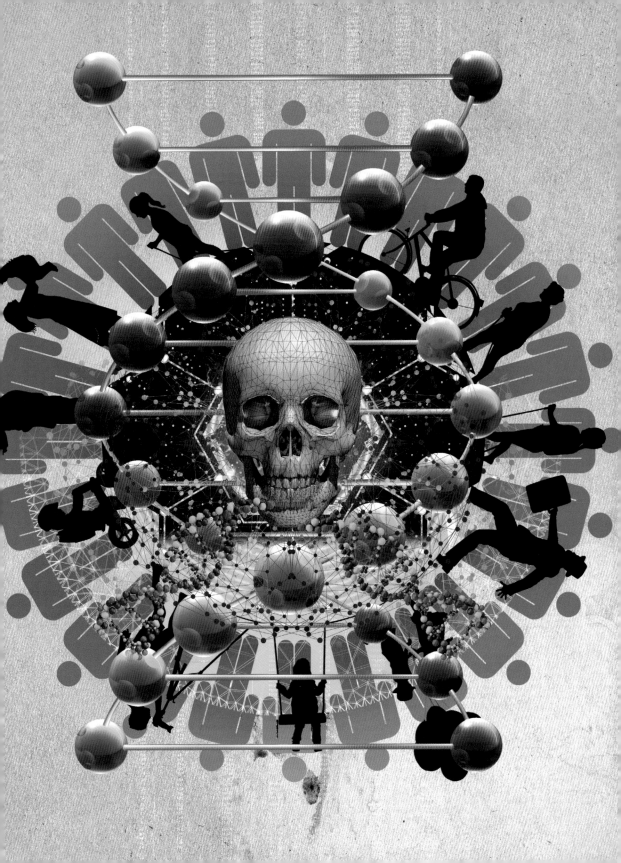

GENES TO ORGANISMS

chromosome Threadlike structure that bears the genes that carry genetic information. Chromosomes are found in the nucleus of eukaryotic cells (those that have a discrete nucleus); they consist mainly of DNA, with some RNA and a core of proteins. A prokaryotic cell (one without a discrete nucleus) has a single chromosome made entirely of DNA.

cloning Asexual reproduction of genetically identical copies of an original organism or cell. Cloning occurs naturally: plants' and animals' body cells are ultimately clones of an original fertilized egg. Cells are also cloned in the laboratory: for example, the nucleus can be removed from an egg and replaced with the nucleus of a cell of the type to be cloned.

DNA Deoxyribonucleic acid, a molecule that carries the coded genetic information that transmits inherited traits. DNA is found in the cells of all prokaryotes and eukaryotes.

Epstein-Barr virus Extremely common virus, also known as the human herpesvirus 4, cause of infectious mononucleosis ('glandular fever') but linked also with types of cancer including Hodgkin's lymphoma (a form of blood cell tumour). It takes its name from its discoverers: Michael A. Epstein and Yvonne Barr.

growth factor Natural substance that stimulates cell growth, healing or reproduction. Examples include some hormones and proteins.

homeostasis Any process by which a biological system maintains its stability. For example, an organism repairs old and damaged cells, using mitosis, to maintain homeostasis.

hormone Signalling molecule, in animals produced in glands and moved in the body's circulatory system to regulate the behaviour and functioning of target cells in organs and tissues. For example, in humans the hormone insulin produced in the pancreas targets the liver and controls blood-sugar levels. Synthetic hormones can be made to produce the same effect as some natural ones.

human papilloma virus (HPV) Common and highly contagious group of viruses that affect the skin and moist membranes in the human body – for example in the anus, genitals, cervix or mouth. Of more than 100 types of HPV, 40 affect the area of the genitals. Some increase the risk of cervical cancer; the HPV vaccine was introduced to counter this risk.

lymphocyte Type of white blood cell with a key role in the human body's immune response. Found in the lymph nodes, spleen and tonsils as well as in the blood circulation, lymphocytes are specialized immune cells with the power to neutralize infection and destroy infected cells.

macrophage Specialized cell in the human immune system found throughout the body. Macrophages engulf and digest damaged or dead cells.

meiosis Cell division in which one parent cell becomes four offspring cells, each one having half the number of chromosomes of the original. Meiosis occurs in preparation for sexual reproduction in eukaryotes.

mitosis Cell duplication in which one cell splits into two daughter cells that are genetically identical – each new cell has the same kind and number of chromosomes.

pathogen Microorganism such as a bacterium or virus that can cause disease.

pluripotent A stem cell capable of becoming any type of cell.

RNA Ribonucleic acid, a compound – found in all living cells – that plays a key role in the synthesis of proteins. In some viruses, RNA rather than DNA functions as the carrier of genetic information.

stem cell Undifferentiated cell that can produce either more stem cells or differentiated (specialized) cells in the heart or spinal nerves, for example.

vaccination Process of inoculating individuals with a vaccine to bring about immunity against a disease. The vaccine stimulates the body's immune response.

CELLS & CELL DIVISION

the 30-second theory

3-SECOND DISSECTION
Mitosis is essential for growth and development, normal homeostasis and damage repair, while meiosis is required for sexual reproduction.

3-MINUTE SYNTHESIS
Cell division can be a risky business, because errors in the process can result in malfunctioning cells or even diseases such as cancer. As a safeguard there are a number of checkpoints at each stage of the cell cycle. The cell will check that all of the DNA is replicated correctly, that the chromosomes are properly separated and that the environment is favourable. If these checks are not passed then cell division will be paused or aborted.

For all forms of life the only way to make a new cell is to duplicate an existing one. Cell division is the process in which one cell duplicates its contents – and divides into two new cells. This process is tightly controlled and consists of a sequence of events known as the cell cycle. In the first phase of the cell cycle the cell grows bigger as it makes copies of its proteins and organelles. In the second phase the cell makes copies of each of its chromosomes. After this the cell enters another period of cell growth and protein synthesis before the final stage sees the chromosomes separated to opposite ends of the cell and the cell splitting itself in half to produce two new daughter cells. This type of cell division is called mitosis and can occur millions of times every second throughout the whole organism as new cells are constantly produced to replace old or damaged ones. Another type of cell division, called meiosis, occurs only in specialized sex cells and is required for sexual reproduction in eukaryotes. For this reason meiosis results in cells (egg or sperm) that only contain one copy of each chromosome, so that after fusion the fertilized egg will contain two copies of chromosomes – one from each parent.

RELATED TOPICS
See also
CELL COMMUNICATION
page 56

CANCER
page 84

CELLULAR SENESCENCE
& DEATH
page 106

3-SECOND BIOGRAPHIES
LELAND HARTWELL
1939–
American cell biologist who introduced the concept of checkpoints and discovered the genes that control the first step of the cell cycle

TIM HUNT
1943–
British biochemist who discovered proteins called cyclins that are essential for control of the cell cycle

30-SECOND TEXT
Phil Dash

Mitosis – one cell splitting into two genetically identical daughters – happens all the time in a healthy organism.

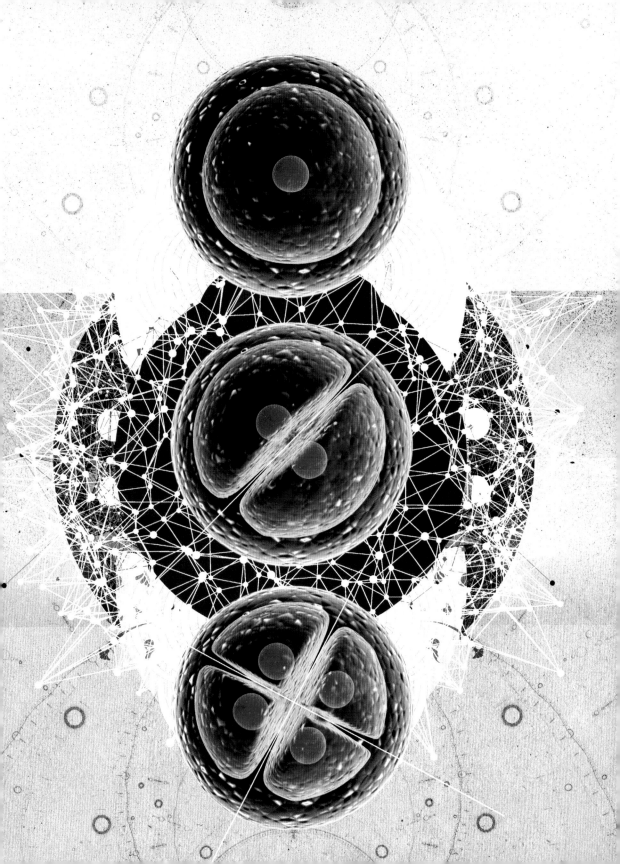

CELL COMMUNICATION

the 30-second theory

3-SECOND DISSECTION
Cells in all multicellular organisms need to be told what to do and so send each other messages that are received by cell surface molecules called receptors.

3-MINUTE SYNTHESIS
Cell communication is so important that when it goes wrong it can lead to a wide range of diseases, the most prominent of which is cancer. In cancer, cells stop listening to signals from other cells and, rather than wait for instructions telling them to divide, they go ahead and divide anyway. This can sometimes be caused by mutations in receptors that mean they are constantly active even in the absence of the appropriate signal.

Cells are constantly bombarded with signals from their environment, whether they are single-celled organisms or cells in tissues within multicellular organisms. These signals provide information about nutrient and oxygen levels, and other features of the cell's environment. In multicellular organisms cells must also coordinate their activities with other cells; hormones, growth factors and other molecules provide instructions to enable this coordination. For example, pancreatic cells release the hormone insulin in response to high blood sugar, which tells other cells in the body to take up sugar from the blood. Other signals may instruct a cell to divide, move, die or change function; they are detected in the target cell by specific molecules on the cell surface called receptors. The receptors are specific to each signal and allow each cell constantly to survey its environment. Receptor binding to the signal molecule initiates a cascade of reactions inside the cell, a process known as signal transduction, which leads to the required change in cell behaviour. For example, in mammals the binding of the signal molecule epidermal growth factor (EGF) to its receptor activates an enzyme cascade that in turn activates the cell cycle, leading to cell duplication.

RELATED TOPICS
See also
CELLS & CELL DIVISION
page 54

IMMUNITY
page 60

CANCER
page 84

3-SECOND BIOGRAPHIES
EARL W. SUTHERLAND
1915–74
American biochemist who showed that hormones work by activating enzymes inside target cells that produce additional signalling molecules called second messengers

MARTIN RODBELL &
ALFRED GILMAN
1925–98 & 1941–
American biochemists who discovered the importance of G-proteins – key molecules that allow signals to be transmitted in cells

30-SECOND TEXT
Phil Dash

How does a cell know when to divide? It receives instructions via receptor molecules on its surface.

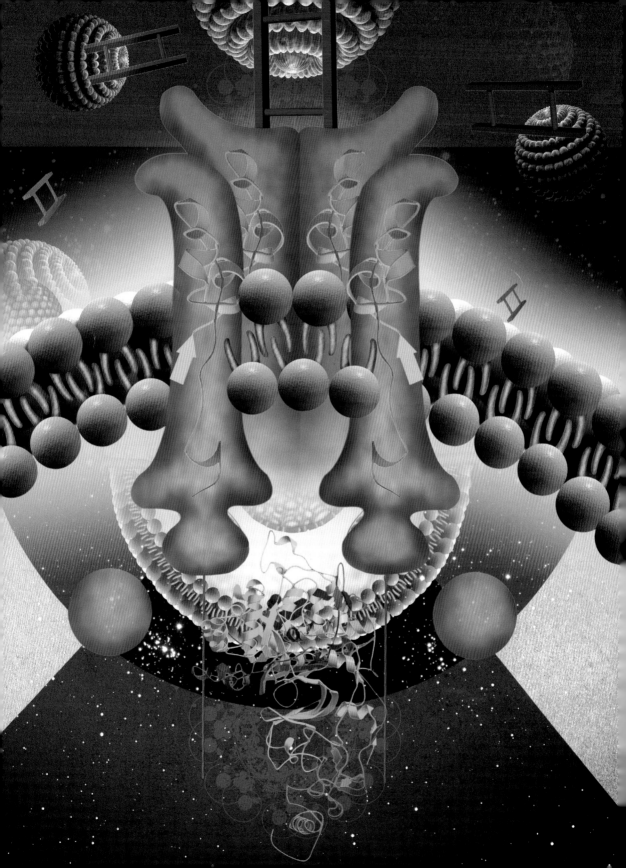

1936
Born in Gelsenkirchen-
Buer, Germany

1960
Receives doctorate in
medicine from the
University of Düsseldorf

1966
Begins work in the virus
laboratory at the
Children's Hospital of
Philadephia

1969
Becomes professor at
the Institute of Virology
of the University of
Würzburg

1972
Becomes professor at the
University of Erlangen-
Nuremberg

1977
Becomes head of the
Department of Virology
and Hygiene at the
University of Freiburg

1983
Identifies first human
papilloma virus DNA in
cervical cancer tumours

1983
Appointed scientific
director of the German
Cancer Research Centre

2004
Awarded the Knight
Commander of the Order
of Merit of the Federal
Republic of Germany

2008
Awarded the Nobel Prize
in Physiology or Medicine

HARALD ZUR HAUSEN

Nobel Prize laureate Harald zur Hausen transformed our understanding of cervical cancer when research on the human papilloma virus (HPV) showed a causal link with cancerous tumours.

As a child in the Second World War, zur Hausen's education was fragmentary, but this did not prevent him starting a medical degree at the University of Bonn, where he studied biology and medicine. Moving on to Hamburg and then Düsseldorf part of the way through his degree, zur Hausen always intended to enter research, but he wanted to qualify as a medical doctor, which meant spending two years as an intern. Immediately after his internship he began work at the University of Düsseldorf's Institute for Microbiology, studying virus-induced chromosome modifications. From there he moved to the virus laboratory of the Children's Hospital in Philadelphia, where he joined virologists Werner and Gertrude Henle. They examined the impact of Epstein-Barr virus on cells, producing the first clear demonstration that a virus could make a healthy cell cancerous.

By his early 40s, zur Hausen was running the virology department in Freiburg, where, with Lutz Glissmann, he isolated the human papilloma virus HPV 6 from genital warts.

With another co-worker – Ethel Michele de Villiers, whom he later married – zur Hausen cloned HPV 6 DNA from genital wart cells, establishing DNA fingerprinting as a potential mechanism for identifying the role of viruses in triggering cancer cells. This research culminated in 1983 and 1984 in the discovery of the DNA of two more human papilloma virus strains (HPV 16 and 18) in cervical cancer tumours, resulting in the establishment of HPV as the cause of the majority of cervical cancer cases.

The HPV discoveries were disputed for some time, as there were conflicting possible causes for cervical cancer, but as further data was accumulated zur Hausen's theory was confirmed. This led to the introduction of an HPV vaccine for young women in 2006, which should produce a major reduction in cervical cancer cases. Zur Hausen took half the Nobel Prize award for 'his discovery of human papilloma viruses causing cervical cancer', with the other half shared between the discoverers of the role of HIV in AIDS. There was some concern that a member of the Nobel Assembly was a director of a drug company involved in the HPV vaccine, but there was no evidence this influenced the outcome and zur Hausen's peers universally endorsed this award for his outstanding work.

Brian Clegg

IMMUNITY

the 30-second theory

The immune system defends
multicellular life from harmful microorganisms. Immunity starts with physical barriers, such as the skin, that prevent the entry of pathogens. Once these defences have been breached our immune system proper comes into play. The initial response to a pathogen typically comes from specialized cells called macrophages. These cells are found throughout the body and possess specialized receptors on their surface that recognize bacterial proteins and viral genetic material. Once detected, the pathogens are quickly engulfed and digested by the macrophages. This is rarely sufficient to wipe out an infection, but it contains the spread of the pathogen until other branches of the immune system can swing into action. The next step is for the macrophages to send out signals to the rest of the immune system, guiding new immune cells such as neutrophils in large numbers to the site of infection. Neutrophils release antimicrobial chemicals and also trigger a state of inflammation, enhancing the immune response further. This type of immune response can be seen in all animals, but another type – based on specialized immune cells called lymphocytes – is found only in vertebrates. Some lymphocytes have the ability to destroy virus-infected cells.

RELATED TOPICS
See also
ORIGINS OF LIFE – VIRUSES
page 14

BACTERIA
page 18

CELL COMMUNICATION
page 56

3-SECOND BIOGRAPHIES
SUSUMU TONEGAWA
1939–
Japanese molecular biologist who discovered the genetic basis for antibody diversity

JULES A. HOFFMANN
1941–
French immunologist who discovered how pathogens are detected by immune cells

BRUCE A. BEUTLER
1957–
American immunologist who discovered how immune detection of microbes functions in mammals

30-SECOND TEXT
Phil Dash

3-SECOND DISSECTION
The immune system contains specialized cells such as macrophages and lymphocytes that can produce an arsenal of chemical and biological weapons to detect and destroy potential pathogens.

3-MINUTE SYNTHESIS
A small fraction of lymphocytes is capable of recognizing specific pathogens. In response to an infection it often takes a few days for these lymphocytes to expand their numbers, but once the infection is beaten many of them will remain as memory cells, capable of responding to a subsequent infection by the same pathogen. This immunological memory is the basis for the protective effect of vaccination.

The immune system unleashes specialized cells to counter harmful invaders.

NEURONS

the 30-second theory

A neuron is a cell in the body specialized to conduct information between the sense organs, the brain and other parts of the body. The information is carried in the form of electrical impulses that 'jump' between adjacent cells via gaps or 'synapses'. Knowledge of the function of synapses has increased vastly in recent years: neurons have adapted processes seen in many other cells to allow the rapid transmission of signals across these small but vital gaps. The signals may pass by direct electrical connection, or through chemical intermediaries called neurotransmitters. A neuron typically receives connections from sense organs or other neurons through short, spine-like connections called dendrites, but transmits information through an axon, which can be hugely elongated. Axons between the spinal cord and the lower limbs can be 1 m (more than 3 ft) in length. Neurons are typically bundled together with other cells and fatty, insulating tissue into nerves. The brain and spinal cord form the central nervous system (CNS). This interfaces through synapses with the peripheral nerve system in the skin, muscles and internal organs. An important part of this is the autonomic nervous system, responsible for unconscious processes such as regulating breathing and heart rate.

3-SECOND DISSECTION
Neurons form the electrical wiring of the brain and body, allowing the various parts of the body to operate in a coordinated way.

3-MINUTE SYNTHESIS
Most cells are responsive to changes in potential difference (voltage) across their membranes; maintaining potential differences across membranes seems to be an essential function of all living cells – even single-celled organisms use electrical potential to govern simple responses to stimuli. Nervous systems evolved in larger, multicellular organisms as part of the division of labour. As some cells became specialized for digestion, secretion or reproduction, others refined and extended these electrical capabilities, evolving into neurons.

RELATED TOPICS
See also
CELL COMMUNICATION
page 56

CONTROVERSY: STEM CELLS
page 68

3-SECOND BIOGRAPHY
SANTIAGO RAMÓN Y CAJAL
1852–1934
Spanish biologist whose drawings of neurons are used to teach neuroscientists

30-SECOND TEXT
Henry Gee

The central nervous system (brain and spinal cord) sends and receives signals in the form of electrical impulses to and from the body's extremities.

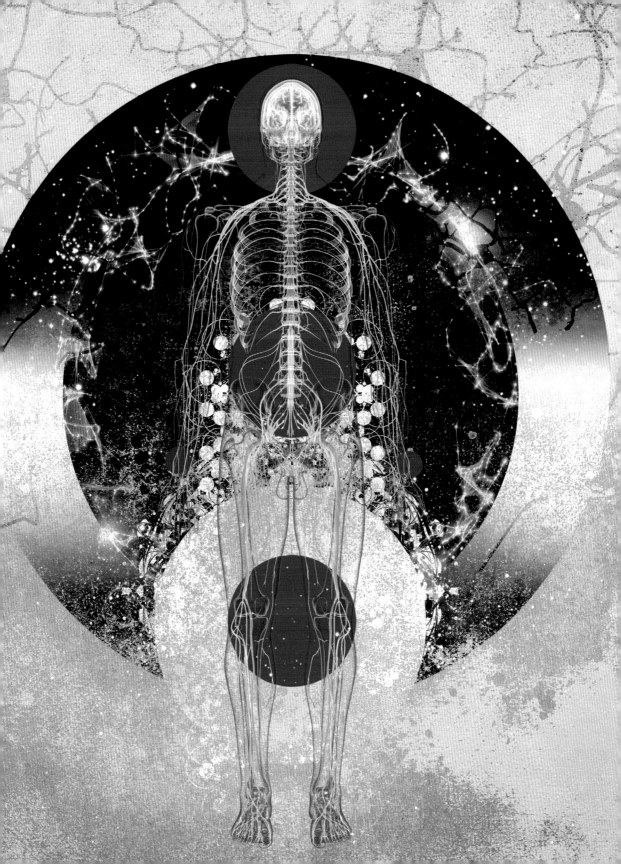

MUSCLES

the 30-second theory

Muscles are about movement: humans walking, fish swimming, birds flying. They also cause movement inside the body, such as the heartbeat, which in humans typically occurs about 60–70 times per minute, supplying oxygenated blood to all organs – including other muscles. For physical movement of the body, muscles are firmly attached to the bones of the skeleton, and hence called 'skeletal' muscles. These are made from bundles of individual muscle cells, each cell containing parallel bundles of protein filaments (myofibrils). Nerve impulses stimulate these cells to contract, exerting kinetic force on the part of the skeleton to which the muscle is attached and creating movement. The other main type of muscle is smooth muscle, which is not connected directly to the skeleton, but is a component of internal organs such as the reproductive system and the intestine. Smooth muscles typically contract in a phased manner, which creates a controlled wave of movement. The muscles that deliver the most force in the body are the uterine muscles, used when mammals give birth to live young. Muscle is also a major source of dietary protein for carnivorous animals, and to a large extent for omnivores (like us); eating roasted skeletal muscle is not something we usually think about when sitting down for dinner.

3-SECOND DISSECTION
Muscles rapidly change chemical energy in an organism into some form of movement.

3-MINUTE SYNTHESIS
There are around 650 skeletal muscles in the human body. The myofibril filaments within muscle cells are mainly made of the proteins actin and myosin. In addition to enabling animals to move about in their environment, muscular movements are needed for sensory evaluation – for example moving and focusing the eye. Muscles in the thorax and larynx enable the generation of sound and the use of language; facial muscles facilitate emotional and non-verbal communication.

RELATED TOPICS
See also
REPRODUCTION: ANIMALS
page 78

METABOLISM
page 100

EXCRETION
page 104

3-SECOND BIOGRAPHY
H.E. HUXLEY & A.F. HUXLEY
1924–2013 & 1917–2012
Brilliant English physiologists who made breakthrough advances in muscle and nerve biology

30-SECOND TEXT
Tim Richardson

We rely on muscles not only to move around but to see, to speak, to smile or frown, to swallow and digest food – and to pump blood around the body.

CIRCULATORY SYSTEM

the 30-second theory

3-SECOND DISSECTION
Circulatory systems across the animal kingdom supply cells with nutrients and oxygen and carry away the waste products of cell function.

3-MINUTE SYNTHESIS
Natural selection has enabled the development of mechanisms that protect the circulatory sytem, preventing the loss of blood. Damaged or broken blood vessels trigger the blood to clot, preventing further leakage, while blood flow in healthy vessels prevents blood clotting. Diseases of blood vessels, often caused by the accumulation of fatty deposits, can result in clotting inside vessels that in humans leads to heart attacks and strokes.

Animal cells require a constant supply of nutrients and oxygen from the environment, and they produce waste products that are released back into the environment. Complex species with many cell layers have developed supply mechanisms that deliver nutrients to cells and remove waste, known as circulatory systems. These have three basic components: fluid (in the case of blood containing many cells), a pump (the heart) and a complex network of tubes or vessels. In insects, spiders and crustaceans the circulation is open and the fluid transported (haemolymph) bathes the tissues and their component cells, and is sucked back into the vessels during heart relaxation. In all vertebrates blood is confined to the vessels and is under pressure that surges as the heart beats. Mammals and birds possess a double circulation system: between the heart and the lungs; and between the heart and the rest of the body. This enables blood to travel to the lungs, release waste carbon dioxide and pick up oxygen, before returning to the heart, which then pumps the fuel-filled blood to the rest of the body through a network of arteries. Arteries divide into smaller and smaller vessels (capillaries) that permeate into all tissues, ensuring that every cell is close to the supply network. Blood returns to the heart through veins, and the cycle begins again.

RELATED TOPICS
See also
RESPIRATION
page 94

METABOLISM
page 100

NUTRITION
page 102

EXCRETION
page 104

3-SECOND BIOGRAPHY
WILLIAM HARVEY
1578–1657
English physician who discovered the organization of the human circulatory system in 1628

30-SECOND TEXT
Jonathan Gibbins

In our double circulation blood pumped from the heart to the lungs returns enriched with oxygen, then is distributed via the arteries before coming back via the veins.

CONTROVERSY
STEM CELLS

the 30-second theory

Stem cells provide the means to

repair damaged tissue such as diseased heart muscle or crushed nerves in the spinal cord. They have enormous potential for regenerative medicine, but are controversial because the most valuable stem cells (pluripotent ones, which can give rise to all cell types) are obtained by destroying human embryos. The central question in the ethical debate is: when is an embryo considered to be a person? It is wrong to destroy a person, even if its stem cells will limit the suffering of another; therefore, if an embryo is a person its stem cells should be off-limits. On the other hand, if personhood is to do with having the basis for a nervous system, and attaining individuality, then before around 14 days an embryo isn't a person. Religious faiths differ over this issue, some stressing the *potentiality* to personhood in any embryo, and therefore set different limits. Laws also vary: in the UK research-only embryos are allowed; it is illegal to implant them in the uterus, so they are therefore not considered potential people. Researchers can also use artificially maintained embryonic stem cell lines to avoid the need for embryos. Induced pluripotent stem cells can now be obtained from adult cells, potentially avoiding embryo use and reducing the ethical problem.

3-SECOND DISSECTION
The use of stem cells is controversial because the cells are obtained from embryos; this dilemma may ease if stem cells can be obtained from adult cells.

3-MINUTE SYNTHESIS
A problem with using embryonic stem cells in regenerative medicine is that the recipient's immune system may reject them. A potential solution is to make a cloned embryo, replacing the nucleus of the donor egg with one from a cell taken from the patient's own body; its stem cells will then be genetically matched.

RELATED TOPICS
See also
CONTROVERSY: GENETIC SCREENING
page 48

DEVELOPMENT: ANIMALS
page 76

CONTROVERSY: GM ORGANISMS
page 88

3-SECOND BIOGRAPHIES
ERNEST MCCULLOCH & JAMES TILL
1926–2011 & 1931–
Canadian scientists who provided key evidence for the existence of stem cells in the early 1960s

JAMES THOMPSON
1958–
American scientist who in 1998 created the first human embryonic stem cell line, and in 2007 obtained human induced pluripotent stem cells

30-SECOND TEXT
Nick Battey

Healing power. Pluripotent stem cells can differentiate into any body cell.

GROWTH & REPRODUCTION

anisogamy Sexual reproduction that combines dissimilar gametes. The opposite – sexual reproduction that fuses similar gametes – is called isogamy.

annual/biennial/perennial Terms that describe the length of a plant's life. Annual plants complete their life cycle, from germination to seed production, in one year; biennials take two years. Perennial plants live for more than two years.

apoptosis Programmed death of cells in an organism. Defects in the genes that promote apoptosis are one cause of cancer – the cells that would normally die do not.

biofilm Self-sufficient community of bacteria in which different species may cooperate, some recycling the wastes of others. Dental plaque is a biofilm.

deciduous A tree or shrub that sheds its leaves annually. An evergreen sheds its leaves gradually throughout the year, generating new ones at the same time.

diploid A nucleus or cell that has two sets of chromosomes, one from each parent.

DNA Deoxyribonucleic acid, a molecule that carries the coded genetic information that transmits inherited traits. DNA is found in the cells of all prokaryotes and eukaryotes.

Dolly the sheep The first mammal to be cloned from an adult cell. In 1996 a team from the Roslin Institute and biotech firm PPL Therapeutics cloned Dolly, a domestic sheep, from a sheep's mammary gland cell by adding new genetic material into a cell from which the original genetic material had been removed.

gamete A haploid (with one set of chromosomes) female or male germ cell that has the capacity to fuse with another of the opposite sex during sexual reproduction to form a zygote. In animals, female gametes are ova (eggs), while male ones are spermatozoa.

gene pool Set of genes in a population. Genetic variation is the variation in genes in a gene pool.

genetically modified An organism that has had foreign genetic material introduced into it to produce desirable traits – for example, resistance to pests in a plant.

haploid A nucleus or a cell having one set of unpaired chromosomes.

intromittent organ External organ used by a male to deliver sperm during sexual reproduction. In mammals this is the penis.

meristem Area at the growing tip of plant shoots and roots where stem cells divide.

natural selection The process by which those organisms that are best adapted to their environment will generally survive and have more offspring. Elucidated by English naturalist Charles Darwin, natural selection is one of the key mechanisms through which evolution works.

photosynthesis The process by which green plants produce their food (sugars and starches) from water and carbon dioxide, giving off oxygen as a byproduct. The process is powered by light energy from the sun, which plants harness through the chlorophyll found in the chloroplasts within their cells.

plasmid Small DNA strand often found in the cytoplasm of a bacterium.

radiotherapy Treatment of disease – typically cancer – using X-rays or other high-energy radiation. Also known as radiation treatment.

RNA Ribonucleic acid, a molecule – found in all living cells – that plays a key role in the synthesis of proteins. In some viruses, RNA rather than DNA functions as the carrier of genetic information.

sepals Part of angiosperms (flowering plants) that can protect the flower in bud and are found outside the petals of the flower. Sepals are typically green and can look rather like leaves.

stamen Male, reproductive organ of a flower. The stamen produces pollen.

transgenic plants Plants containing genetic material that has been transferred from another organism.

tumour Swelling caused by abnormal tissue growth. Cancerous tumours, caused by abnormal cell growth and with the potential to spread to other parts of the body, are termed malignant. Noncancerous tumours are termed benign.

zygote A diploid cell (one with two sets of chromosomes), formed by the fusion of two gametes – e.g., an egg that has been fertilized by a sperm.

DEVELOPMENT & REPRODUCTION: BACTERIA

the 30-second theory

Bacteria are single-celled

organisms, surrounded by tough walls. They multiply by splitting into two new, distinct and genetically identical cells. The dividing time of bacteria varies enormously. *Escherichia coli*, the common gut bacterium, has a 'doubling time' of less than 20 minutes. The infectious agent of tuberculosis, *Mycobacterium tuberculosis*, can take up to 16 hours. Because bacteria are only susceptible to antibiotics when they are dividing, this means that slow-dividing bacteria are hardest to kill. Typically, bacteria on a suitable food source keep on dividing until they exhaust their resources, after which they stop dividing and may die. Some bacteria, however, form stable associations with other bacterial species to form a 'biofilm', a kind of self-sufficient ecological community in which different species recycle the wastes of others. Biofilms are found in, for example, dental plaque, and also the lungs of patients with cystic fibrosis, and are very hard to remove. When confronted by hard times, some kinds of bacteria such as *Clostridium tetani*, which causes tetanus, form a different kind of structure – an inert and resistant state called a spore. Because it forms a distinct kind of cell, spore formation is just about the only process that counts as 'development' in bacteria.

3-SECOND DISSECTION
Like most single-celled organisms, bacteria multiply by dividing; although their potential for individual development is limited, communication and genetic exchange between individuals causes differentiation within communities.

3-MINUTE SYNTHESIS
Bacterial life isn't completely sexless. Sometimes bacteria extend tubes called 'pili' to one another, through which they exchange DNA. In addition to their chromosomes, bacteria may host smaller pieces of DNA called plasmids. These are important as they may contain genes that confer antibiotic resistance or other traits, like the ability to fix nitrogen. In general, bacteria are liberal with their DNA, scavenging it from the environment and donating it to one another.

RELATED TOPICS
See also
ARCHAEA
page 16

BACTERIA
page 18

MUTUALISMS
page 122

3-SECOND BIOGRAPHIES
OSWALD AVERY
1877–1955
American biologist who in 1944, with colleagues Colin MacLeod and Maclyn McCarty, showed that live bacteria scavenged DNA from dead cells, proving that DNA was the hereditary material

STANLEY N. COHEN
1935–
American geneticist who, with Herb Boyer and Paul Berg, discovered how to use bacterial plasmids to transfer DNA from one organism to another, so inventing genetic engineering

30-SECOND TEXT
Henry Gee

Bacteria such as E. coli can divide at speed until they run out of resources.

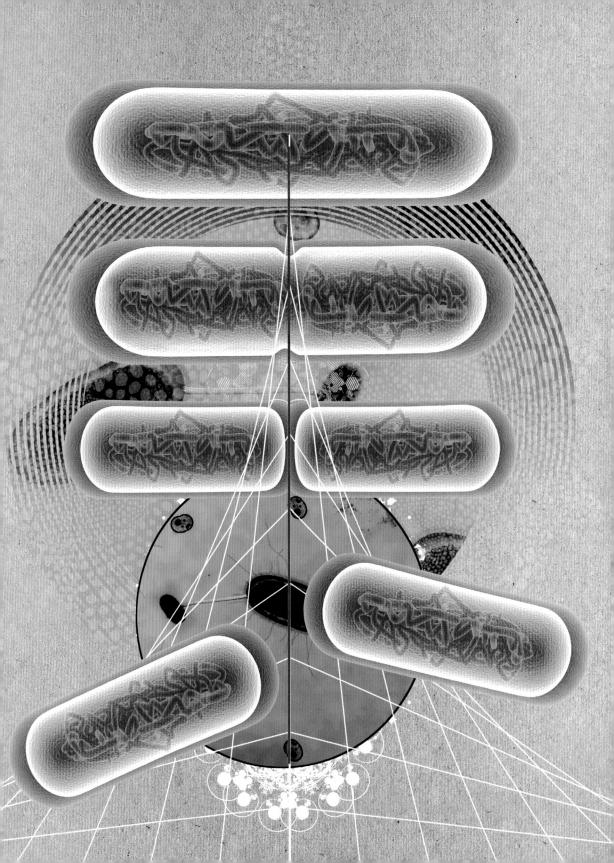

DEVELOPMENT: ANIMALS

the 30-second theory

3-SECOND DISSECTION
Animal development is a process by which egg and sperm cells conspire to create and disperse more egg and sperm cells.

3-MINUTE SYNTHESIS
In the 1980s, biologists found that peculiarities long known to occur in fruit flies – such as extra pairs of wings or legs growing in place of antennae – were mutations in a group of genes (the Hox genes) that determine the spatial order in which structures develop in the body. Hox genes were found to do much the same things in all animals, including humans. This finding revolutionized our understanding of development.

All animals develop from a single cell, the zygote, created by the fusion of an egg and a sperm. At first, the zygote divides into a ball of smaller cells, without growing overall. The developmental pathway followed by these cells depends on the distribution of yolk derived originally from the maternal egg cell. Cells that contain yolk are larger, divide more slowly, and become the gut of the adult animal. Cells without yolk are smaller, divide more quickly and become the skin, nerves and outer parts of the creature. Typically, the ball hollows out to become a 'blastula', with a central cavity (the 'blastocoel'), then collapses like a squashed football to produce a 'gastrula', made up of two layers of cells and an opening at one end. In some animals this opening becomes the mouth. In others – notably vertebrates, including humans – it becomes the anus, the mouth developing from a secondary opening. In most animals, cells from the inner and outer layers accumulate in the space between, lining the blastocoel to produce the adult body cavity, and internal organs such as muscles and blood vessels. Some creatures, such as roundworms and tunicates, have highly determinate development – that is, the lineage of each cell in the adult can be traced back to the zygote.

RELATED TOPICS
See also
ANIMALS
page 28

REPRODUCTION: ANIMALS
page 78

3-SECOND BIOGRAPHIES
WILHELM ROUX
1850–1924
German embryologist who studied under Ernst Haeckel and perhaps did the most to create the study of embryology and development as we know it

CHRISTIANE NÜSSLEIN-VOLHARD
1942–
German biologist who, with Edward Lewis and Eric Wieschaus, was awarded the 1995 Nobel Prize for his work on Hox genes in the fruit fly

30-SECOND TEXT
Henry Gee

Research into fruit flies' hox genes helped us understand more clearly how gene mutations lead to changes in body structures.

REPRODUCTION: ANIMALS

the 30-second theory

Reproduction provides the

opportunity for genes to continue and perhaps spread even if their living home dies. It allows for genetic variation, grist to the mill of natural selection. Individual animals of sexually reproducing species produce haploid gametes (spermatozoa or ova) and then devise ways to bring them together to create a new diploid being. The diversity of life is matched by the diversity in ways of achieving this continuity. Most animals shed gametes into the water in which they live, and fertilization is external. Many, though, maximize the chances of fertilization by devising strategies of internal fertilization, usually involving the insertion of a male intromittent organ, such as a penis, into the female. Even when fertilization is internal, the zygotes often develop externally, as eggs. Because ova are large and few, while sperm are small and many, and because females tend to invest more energy than males in their offspring, it is in the female's interest to be much choosier about the male with whom she mates than vice versa. The courtship strategies and different styles of parental care seen in the animal kingdom stem from this fundamental sexual conflict. This in turn stems from the unequal size of the gametes (anisogamy) – the reasons for which are still debated by scientists.

3-SECOND DISSECTION
Animal reproduction is hugely varied but usually boils down to a story of male-meets-female, and how they contribute to the next generation.

3-MINUTE SYNTHESIS
Many animals and plants reproduce asexually, by (for example) dividing, or budding off, copies of themselves. In view of this, the origin of sexual reproduction seems a puzzle: a mate will pass on half rather than all its genes to the next generation. Many explanations for sex have been proposed. One is that sex shuffles genes, keeping the gene pool healthy and varied, which is an important hedge against disease, parasitism and unforeseen environmental circumstances.

RELATED TOPICS
See also
DEVELOPMENT: ANIMALS
page 76

DEVELOPMENT: PLANTS
page 80

SEXUAL SELECTION
page 116

3-SECOND BIOGRAPHY
AUGUST WEISMANN
1834–1914
German biologist who first realized that in multicellular organisms inheritance is governed by special sex cells or 'germ plasm'

30-SECOND TEXT
Henry Gee

Asexual or sexual, the egg fertilized within the body or outside it . . . there is wide variety in animal reproduction.

DEVELOPMENT: PLANTS

the 30-second theory

Unlike animals, adult plants spend their entire lives in one spot, but compensate by immense variability in body shape. Whereas an animal is compact, usually with a single head, a fixed number of limbs and so on, plants can vary enormously as regards numbers of leaves, flowers, roots and branches. This makes plant development fundamentally different from animal development. Even so, plants do have standard, repeating structures: leaves and stem tissue make up a shoot and are eventually followed by flowers; roots grow according to a fixed pattern. Each root or shoot develops from a microscopic region called a meristem at the growing tip, consisting of a group of actively dividing stem cells. This means that shoots and roots grow from the tip. The environment plays an important part in plant development. Sensitivity to gravity ensures that roots grow downwards and shoots grow upwards; the need to harvest sunshine by photosynthesis means that plants bend towards the light. In temperate regions plant development also varies markedly during the year. Deciduous trees respond to the shortening days and cooler temperatures of autumn by withdrawing valuable materials from their leaves. The green chlorophyll pigment is recycled, revealing vivid autumn colours.

Goethe expressed his theory about plants in his 1790 book, Metamorphosis of Plants.

REPRODUCTION: PLANTS

the 30-second theory

RELATED TOPICS
See also
DEVELOPMENT: ANIMALS
page 76

REPRODUCTION: ANIMALS
page 78

DEVELOPMENT: PLANTS
page 80

3-SECOND BIOGRAPHIES
NEHEMIAH GREW
1641–1712
English botanist, author of *The Anatomy of Plants* (1682), who realized that pollen was for sex

WILHELM HOFMEISTER
1824–77
German botanist who discovered alternation of generations in plants

30-SECOND TEXT
Henry Gee

3-SECOND DISSECTION
The evolution of plant development is a shifting balance between the gametophyte and sporophyte generations, with the latter dominant in higher plants.

3-MINUTE SYNTHESIS
Plants can reproduce asexually. As every gardener knows, it is often possible to produce new plants from a piece taken from an existing plant. In animals, such regenerative powers are confined to very simple creatures. The fact that even the most evolutionarily sophisticated plants are capable of such feats is probably a reflection of their lack of cell and tissue specialization, which allows a ready ability to regenerate.

Like animals, plants usually

consist of many cells that contain two copies of the genetic material (that is, are diploid) except for the sex cells, or gametes, which contain one copy each (haploid). Fusion of the two in sexual reproduction restores the diploid number. However, this 'alternation of generations' is more prominent in plants than in animals, where the haploid generation is confined to the gametes. In mosses and liverworts, the haploid generation or 'gametophyte' constitutes the main body of the plant. Female gametophytes produce large gametes, which are fertilized by the much smaller spermatozoa, produced by the male gametophyte. The diploid generation, or sporophyte, is often quite small, and produces spores. In ferns, like bracken, the sporophyte is large, the gametophyte small. In gymnosperms (such as conifers) and angiosperms (flowering plants) the sporophyte is dominant, the gametophyte more or less confined to the gametes themselves. In simple plants sperm swim through water (explaining why mosses and liverworts live in damp places) but in complex plants such as gymnosperms and angiosperms the cells that will give rise to sperm are found within pollen, and therefore often reach the female gametes via pollinators or the wind; they produce distinctive diploid propagules – seeds.

Mosses and liverworts need water for the transfer of reproductive cells; flowering plants and conifers rely on pollen moved by wind or a pollinating insect.

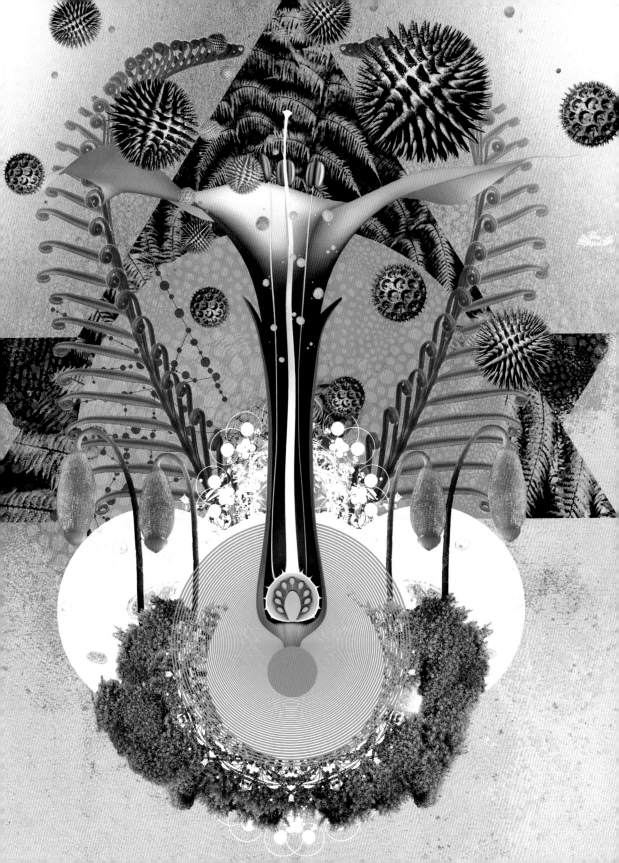

CANCER

the 30-second theory

Cancer is a collection of diseases

caused by uncontrolled cell growth triggered by mutations in the DNA of the cell. This results in the formation of a tumour that if left untreated can eventually lead to cancerous cells entering the blood stream and growing in other parts of the body – a process known as metastasis. The World Health Organization (WHO) identifies cancer as a leading cause of disease and death worldwide. Cancer is typically a disease of older people because most cancers are caused by a build-up of mutations in the DNA that make one cell or a few cells divide out of control. Most of these changes occur by chance over the course of a person's life, but there are lifestyle choices that increase its likelihood, such as smoking, obesity, diet, alcohol consumption and exposure to ultraviolet rays from the sun; it's thought that more than 30 per cent of cancers could be prevented by modifying or avoiding key risk factors. There are many mutational steps required to make normal cells cancerous. These include faulty cell signalling and response, increased cell mobility and survival in the blood stream, and cell immortality (when cells carry on living – and dividing – beyond their natural lifespan). Cancer research tends to focus on these characteristics as targets for treatment because they are shared across cancer types.

RELATED TOPIC
See also
CELLS & CELL DIVISION
page 54

3-SECOND BIOGRAPHIES
MARIE SKŁODOWSKA-CURIE
1867–1934
Polish-born physicist whose work on radioactivity was applied to medicine as radiotherapy

LEOPOLD FREUND
1868–1943
Austrian-Jewish scientist who developed radiotherapy – a cancer treatment still used today

30-SECOND TEXT
Tiffany Taylor

3-SECOND DISSECTION
Cancers are your body's own cells that have mutated to become immortal. All cancers share common characteristics that allow them to grow and spread.

3-MINUTE SYNTHESIS
Cancer incidence in humans is far lower than expected, given the number of cells in our bodies and average life expectancy; this is also true for organisms across the animal kingdom. Animals have evolved tumour suppression genes that stop cells from dividing in an uncontrolled way and/or promote apoptosis (programmed cell suicide). Defects in these genes often lead to cancers. For example, the tumour suppressor protein p53 is mutated in more than 50 per cent of human cancers.

Many different types of mutation can cause cells to grow uncontrollably.

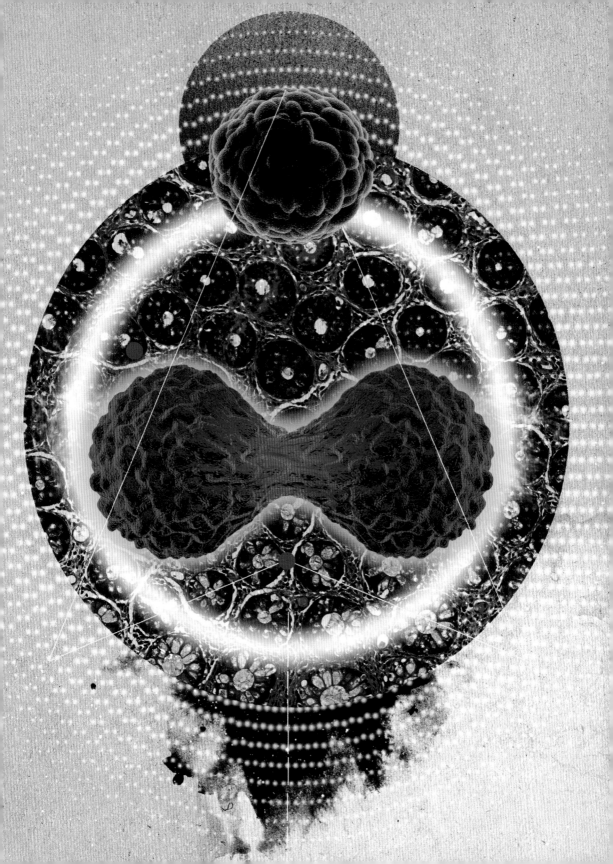

1948
Born in Snug, Tasmania to medical parents

1974
Gains doctorate from University of Cambridge

1975
Marries John Sedat

1975
Takes up postdoctoral research at Yale

1975
Begins to focus on telomeres

1977
Moves to University of California, San Francisco

1978
Moves to University of California, Berkeley, as an assistant professor

1986
Becomes full professor at Berkeley

1986
Her son Benjamin David is born

1990
Moves with her laboratory to University of California, San Francisco's Department of Microbiology and Immunology

1998
Serves as the president of the American Society for Cell Biology

2001-03
Serves on the US President's Council on Bioethics

2009
Awarded the Nobel Prize in Physiology or Medicine

2012
Receives the American Institute of Chemists' Gold Medal

ELIZABETH BLACKBURN

Tasmanian-born Elizabeth

Blackburn has spent a career in molecular biology, where she won the Nobel Prize for her work on the function of telomeres, the 'tags' on the ends of chromosomes.

Born in a small seaside town in the Australian state of Tasmania, Blackburn developed an early interest in animals, collecting tadpoles and living in a house with a menagerie of pets that ranged from canaries to cats and a dog. This interest continued as the family moved first to the larger town of Launceston, then to the city of Melbourne, where Blackburn took her undergraduate degree in biochemistry. After obtaining an MSc at Melbourne, she was awarded a PhD by the University of Cambridge, working under Fred Sanger at the Laboratory of Molecular Biology on the DNA sequence of bacteriophage (a virus that targets bacteria).

After marrying John Sedat, who was about to leave for Yale University, Blackburn changed her plans to undertake postdoctoral work at the University of California, San Francisco, instead obtaining a post at Yale. Here she began to develop the means to sequence the DNA at the ends of short 'minichromosomes' found in a single-celled protozoan. The work on these terminal parts of the DNA, called telomeres from the Greek for 'end' and 'form', continued at the University of California, San Francisco, where both Blackburn and her husband finally obtained posts. It was here that – jointly with Carol Greider – she discovered the enzyme telomerase, which has the role of adding additional sections to the ends of telomeres, reversing their natural tendency to be shortened. As Blackburn has put it: 'Together with colleagues … I was able to address the wondrous biological systems comprised of telomeres and telomerase.'

In the twenty-first century, Blackburn has been more involved in the social and ethical impacts of genetics and molecular biology, particularly during her time on George W. Bush's President's Council on Bioethics, where she felt that she could help bring a strong base of scientific evidence to debates that were politically charged. She was dismissed from the Council after two years when she expressed views differing considerably from those of the administration and the Council's chair, resulting in a wave of public support for her position. In 2009, Blackburn won the Nobel Prize, split evenly with Greider and Jack Szostack, for 'the discovery of how chromosomes are protected by telomeres and the enzyme telomerase'.

Brian Clegg

CONTROVERSY
GM ORGANISMS

the 30-second theory

'GM' stands for genetically modified and invites controversy because it means everything and nothing. Since the beginning of settled agriculture about 12,000 years ago we humans have been selecting those strains of wild species that best suit our purpose. This is genetic modification – and it created domesticates like bread wheat, farm animals and cats and dogs. In modern animal science 'GM' brings to mind Dolly the sheep, a cloned animal produced by fusing a cell from a sheep's udder with an egg cell with its nucleus removed. The resulting egg cell (with the nucleus of the udder cell) developed into an embryo, which became Dolly. Using this technique, valuable transgenic animals (animals with a foreign gene inserted) can be multiplied, creating identical copies. In modern crop science 'GM' implies that a new variety such as herbicide-resistant maize contains a gene or genes inserted by non-conventional breeding methods, such as firing foreign DNA into a target plant cell using a gene gun. Concerns have arisen about crossing organism boundaries – for example, putting fish genes into plants. But the human need for food means we have to cultivate crops in more diverse environments; here, if nowhere else, the ability to create special plants using 'GM' is likely to be valuable.

GM maize and Dolly the sheep are trailblazers of genetic modification, although neither is free of controversy.

ENERGY & NUTRITION

amino acids Water-soluble organic compounds that are constituent parts of proteins. Of around 24 amino acids involved in the making of proteins, 10 cannot be made by the human body and so must be included the diet: they are called essential amino acids.

biodiversity The variety of plants, animals and microorganisms in a habitat. In general usage, biodiversity means the variety of species in a particular habitat (for example, in Antarctica or the Amazon rainforest) or on Earth.

carbohydrates Organic compounds containing carbon, hydrogen and oxygen. Smaller molecules include sugars (saccharides) such as glucose and sucrose; larger ones (polysaccharides) include starch and cellulose. In animals they are broken down to release energy.

cells Smallest units of an organism, typically but not always consisting of a nucleus and cytoplasm encircled by a membrane.

chloroplast A plastid (type of organelle) found in the cells of green plants, in which photosynthesis occurs.

countercurrent multiplier mechanism Process by which urine is concentrated in the kidneys of mammals.

cyanobacteria Water-dwelling single-celled organisms (related to bacteria) that derive energy through photosynthesis. Also known as blue-green algae, they are the earliest known lifeform on Earth – their fossil record in Australia dates to 3.5 billion years ago.

cytoplasm The total content of a cell held within the outer cell membrane.

dwarf wheat A variety of wheat with shorter, thicker stems and higher potential yield.

electron transport chain Series of reactions in which electrons are transferred between compounds – for example, as part of photosynthesis in the chloroplasts of plants, or respiration in the mitochondria of animals and plants.

food security Situation in which a human population has access at all times to enough affordable, safe and nutritious food for people to live an active and healthy life. When this situation is not attained, analysts speak of food insecurity.

genes Units of heredity, located on a chromosome. Genes consist of DNA, except in some viruses where they are made of RNA. Particular genes control specific processes – for example, one gene may control apoptosis (cell suicide) in a range of organisms.

glycolysis Series of chemical reactions in which glucose is broken down by enzymes, with the release of energy.

metabolic pathway Sequence of reactions involved in the breakdown or synthesis of compounds during metabolism.

metabolites Molecules needed for, or formed by, cell metabolism.

metabolome Complete set of small molecule chemicals in an organism. Variations in the metabolome may indicate disease.

molecules Bonded atoms, the smallest part of a chemical compound that can be engaged in a chemical reaction.

organelle A compartment or structure within a cell.

photosynthesis The process by which green plants produce their food (sugars and starches) from water and carbon dioxide, giving off oxygen as a byproduct. The process is powered by light energy from the sun, which plants absorb in the chlorophyll found in the chloroplasts within their cells.

symbiosis and endosymbiosis Symbiosis is reciprocal action involving two organisms that live close to one another and interact, possibly to their mutual benefit. Endosymbiosis occurs when one of the symbiotic organisms is inside the other. For example, biologists believe that chloroplasts (organelles in the cells of green plants) evolved from organisms similar to cyanobacteria by endosymbiosis.

vitamins Organic compounds needed by organisms for normal, healthy life. They cannot be made by the human body so are essential in the diet. Vitamin deficiency (when the organism is lacking the vitamins it needs) can lead to serious diseases in people: for example, scurvy (vitamin C deficiency) and rickets (vitamin D deficiency).

RESPIRATION

the 30-second theory

3-SECOND DISSECTION
Energy from food is released in a series of chemical reactions known as respiration, leading to the production of energy storage molecule ATP.

3-MINUTE SYNTHESIS
How does oxidative phosphorylation in the mitochondria produce so much ATP? When energy is transported from one carrier to another in the mitochondria, a small amount of energy is released at each step and is used to pump protons across the membrane. These protons then move back across the membrane by entering special channels on a protein called ATP synthase. As protons move into these channels they cause ATP synthase to rotate, providing the energy to make ATP.

Cells need a constant supply of energy to stay alive. In animals this comes from food, stored in molecules of carbohydrates, fats and proteins. It is not released all at once during digestion as this would be inefficient; instead the molecules pass through a complex series of steps, each one a carefully controlled chemical reaction, in a process called respiration. The end result is that the energy in food is converted into ATP (adenosine triphosphate), which is used as a cellular power source. In the first stage, complex molecules are broken down to simple sugars. Then respiration takes place in one of two places in the cell. Glycolysis occurs in the cytosol (fluid within the cytoplasm) and oxidizes the sugars, in the process generating two ATP molecules per sugar molecule. A more efficient process, known as oxidative phosphorylation, occurs in mitochondria and can produce more than 30 additional molecules of ATP for each sugar molecule released from food. This process is heavily dependent on oxygen, however, which is converted to water. This is the main use for oxygen in animals: around 90 per cent of the oxygen that animals breathe is used in the process of respiration.

RELATED TOPICS
See also
PHOTOSYNTHESIS
page 98

METABOLISM
page 100

NUTRITION
page 102

3-SECOND BIOGRAPHIES
HANS KREBS
1900–81
German-born British biochemist who identified the biochemical pathways involved in cellular respiration

PETER MITCHELL
1920–92
British biochemist whose theory of chemiosmosis explained how mitochondria generate energy

30-SECOND TEXT
Phil Dash

In respiration food molecules are broken down, releasing energy that is stored in ATP to power cells. Virtually every cellular process requires ATP.

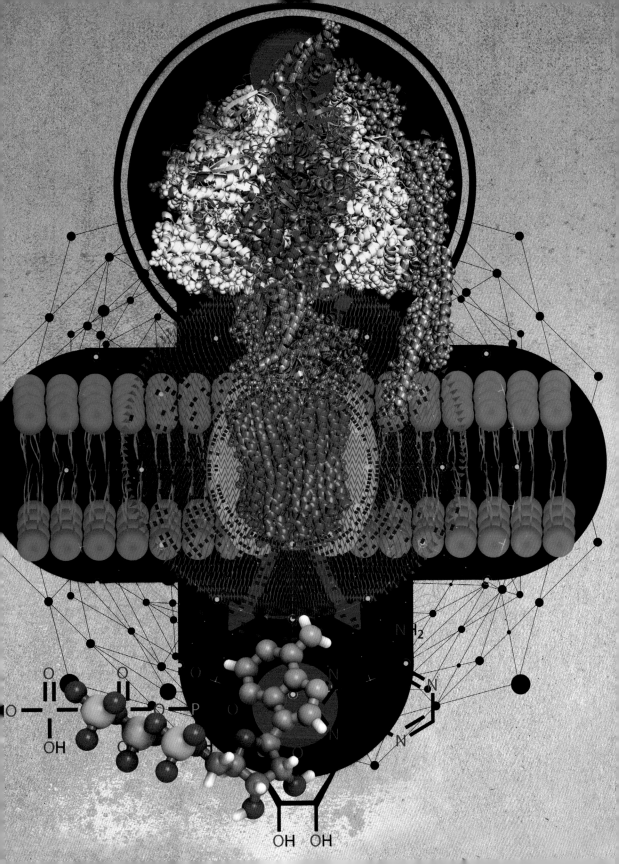

1914
Born near Cresco, Iowa, to Henry Oliver and Clara Borlaug

1933
Enrols at University of Minnesota

1942
Obtains PhD in plant pathology, University of Minnesota

1944–64
Develops high-yielding, disease-resistant, semi-dwarf wheat as head of the Cooperative Wheat Research and Production Program, Mexico

1964–79
Director of the International Wheat Improvement Program, International Maize and Wheat Improvement Centre (CIMMYT), Mexico

1965
Introduces semi-dwarf wheat to India and Pakistan

1968
Work in Mexico, India and Pakistan becomes a 'Green Revolution'

1970
Receives Nobel Peace Prize for contributions to world peace through increasing food production

1977
Receives Presidential Medal of Freedom (USA)

1984–2009
Distinguished professor of international agriculture, Texas A&M University

1986
Establishes World Food Prize

2006
Receives Congressional Gold Medal (USA)

2009
Dies in Dallas, Texas

NORMAN BORLAUG

Norman Ernest Borlaug,

celebrated as 'father of the green revolution', was born and raised in a rural farming community in Iowa, USA. He worked on the family farm until the age of 19, when support from a Depression-era educational programme allowed him to enrol at the University of Minnesota, where he obtained his BS in 1937, his MS in 1940 and his PhD in plant pathology in 1942. After university he worked for DuPont in Wilmington, Delaware as a microbiologist. Then in July 1944 he accepted a position as head of the newly created Cooperative Wheat Research and Production Program, a joint venture between the Rockefeller Foundation and the Mexican government.

In this role he bred a series of high-yielding, disease-resistant, semi-dwarf wheat varieties which, combined with modern agricultural techniques, increased Mexican wheat yields threefold per hectare between 1944 and 1963, and enabled Mexico to become self-sufficient in wheat production. Officials quickly realized that this achievement could be translated to other nations and other crops. In 1964, Borlaug was appointed director of the International Wheat Improvement Program at the International Maize and Wheat Improvement Centre (CIMMYT) at El Batán, Mexico.

In 1965, following several years of field trials, Borlaug's wheat varieties were introduced to India and Pakistan. Even in the first season they out-yielded traditional varieties. These new varieties increased wheat yields by 40–60 per cent per hectare between 1963 and 1970, and saved millions of people from starvation. The work was hailed a 'green revolution' by William Guad, director of USAID, and Borlaug's contribution to global food security was recognized by the award of the Nobel Peace Prize in 1970.

In 1986, Borlaug established the World Food Prize, which recognizes individuals who have improved the quality, quantity or availability of food in the world. Borlaug retired from CIMMYT in 1983, and in 1984 he became distinguished professor at Texas A&M University, where he worked until his death in 2009. In his long career Borlaug received the Presidential Medal of Freedom and the Congressional Gold Medal (the highest civilian awards of the USA), the Aztec Eagle (the highest decoration awarded to foreigners by Mexico) and the Hilal-i-Imtiaz and Padma Vibhushan (the second highest civilian honours of Pakistan and India). A statue of Borlaug was unveiled at the US Capitol in March 2014 to celebrate the centenary of the birth of this remarkable humanitarian.

Philip J. White

PHOTOSYNTHESIS

the 30-second theory

3-SECOND DISSECTION
Photosynthesis converts carbon dioxide (CO_2) and water into carbohydrates using light energy.

3-MINUTE SYNTHESIS
Photosynthesis removes CO_2 from the atmosphere. Between 40 and 70 per cent of all photosynthesis occurs in marine ecosystems. This is limited by the iron needed for the electron transport chain in photosynthesis. Scientists have tried seeding seas with iron to increase global photosynthesis and CO_2 fixation – and slow down climate change.

Photosynthesis produces all the organic carbon for life. It is performed by bacteria (purple bacteria, green sulphur bacteria, cyanobacteria), algae and plants. In algae and plants, photosynthesis occurs in organelles called chloroplasts, which evolved from organisms resembling cyanobacteria by endosymbiosis. Photosynthesis comprises light-dependent and light-independent reactions. In plants, the light-dependent reactions occur in the membranes of chloroplasts. Light energy is absorbed by pigments (primarily chlorophylls) and initiates a sequence of reactions catalyzed by an electron transport chain that converts water to oxygen and produces two energy-rich compounds called adenosine triphosphate (ATP) and nicotinamide adenine dinucleotide phosphate (NADPH). The light-independent reactions, called the Calvin-Benson-Bassham (CBB) cycle, use these compounds to fix CO_2 into carbohydrates within the chloroplast. The product of the CBB cycle is a three-carbon (C_3) molecule. The CBB cycle occurs in leaf mesophyll cells of C_3 plants and uses the enzyme RuBisCO, which is probably the most abundant protein on Earth. Some plants, known as C_4 plants, separate the assimilation of CO_2 from light-dependent reactions.

RELATED TOPICS
See also
LYNN MARGULIS
page 20

METABOLISM
page 100

3-SECOND BIOGRAPHIES
ROBERT HILL
1899–1991
English biochemist who characterized the light-dependent reactions of photosynthesis

MELVIN CALVIN
1911–97
American biochemist awarded the Nobel Prize in Chemistry in 1961 for the discovery of the Calvin-Benson-Bassham cycle

30-SECOND TEXT
Philip J. White

Through photosynthesis light energy is converted into the chemical energy needed for life.

METABOLISM

the 30-second theory

The term 'metabolism'

encapsulates all of the biochemical processes that go on in an organism. Metabolic activity that leads to the synthesis of new cells and their components is referred to as anabolic metabolism. In plants this is driven by photosynthesis, which uses light energy from the sun to convert CO_2 from the air around us into energy-rich sugars. In animals anabolic metabolism begins with eating, drinking and breathing, because these activities provide the cells of the body with the materials they need to generate energy and synthesize new molecules such as proteins. Metabolic activity resulting in the breakdown of molecules and cells is catabolic metabolism, which generates waste compounds excreted by the organism. CO_2, a by-product of respiration, is carried by the blood and released from the body through the lungs; urea, which is largely generated as a metabolic waste product in the liver, is removed by the kidneys. Hormones and drugs that influence metabolism can have beneficial medical applications, for example for treating chronic wasting conditions. However, drugs that influence metabolism are most commonly discussed in the context of performance enhancement for athletes – for example, anabolic steroids.

RELATED TOPICS
See also
RESPIRATION
page 94

PHOTOSYNTHESIS
page 98

EXCRETION
page 104

3-SECOND BIOGRAPHIES
HANS KREBS
1900–81
German-born scientist whose work on metabolic pathways is the basis of our understanding of metabolism

KENNETH BLAXTER
1919–91
English animal nutritionist renowned for his whole animal studies with ruminants

30-SECOND TEXT
Tim Richardson

Anabolic steroids affect our metabolism by increasing the synthesis of proteins within skeletal muscle cells.

NUTRITION

the 30-second theory

A substance is an essential
nutrient if an organism cannot synthesize
enough of it to maintain normal bodily functions
and instead has to obtain it from an external
source. The science of nutrition describes how
essential nutrients are acquired and used. The
list of essential nutrients differs from one
organism to another. Essential nutrients for
humans include: carbohydrates, which are the
main source of energy (calories) and fibre; at
least two specific polyunsaturated fatty acids;
nine amino acids, which can be obtained from
protein; 13 vitamins; and 22 mineral elements.
The main sources of most of these essential
nutrients are edible crops. In terms of calories,
there appears to be sufficient food for
humanity: whilst one-sixth of the global
population is starving, another one-sixth is
obese. However, the diets of most people lack
sufficient vitamins, particularly vitamin A, and
mineral elements, such as iron, zinc, calcium,
iodine and selenium. People who don't get
enough nutrients risk serious diseases such as
scurvy (vitamin C deficiency), beriberi (vitamin
B1 deficiency), rickets (vitamin D or calcium
deficiency), anaemia (iron deficiency) and goitre
(iodine deficiency). Mineral malnutrition is often
associated with growing crops on soils low in
essential minerals.

3-SECOND DISSECTION
Nutrition describes how an
organism gets and uses
essential nutrients – the
elements and organic
compounds it needs for
optimal functioning that
must be obtained from an
external source.

3-MINUTE SYNTHESIS
Edible crops are the
primary source of many of
the essential nutrients in
human diets. Plants can
make all their organic
compounds from inorganic
compounds and elements
in the environment. They
assimilate carbon, oxygen
and hydrogen through
photosynthesis. Their roots
obtain essential mineral
elements from the soil.
Thus, photosynthesis and
the mining of soils by plant
roots ultimately provide
the nutrients that are
essential to humans.

RELATED TOPICS
See also
PHOTOSYNTHESIS
page 98

METABOLISM
page 100

FOOD WEBS
page 138

3-SECOND BIOGRAPHIES
JAMES LIND
1716–94
Scottish physician who first
demonstrated that scurvy
could be remedied by eating
citrus fruit

ELSIE WIDDOWSON
1906–2000
English dietician who, with
Robert McCance, co-authored
*The Chemical Composition of
Foods*, which informed modern
nutritional thinking

30-SECOND TEXT
Philip J. White

*People risk serious
disease if their diet
doesn't include enough
vitamins and minerals.*

EXCRETION

the 30-second theory

3-SECOND DISSECTION
Excretion is all about eliminating wastes left over from metabolism, especially nitrogen from broken-down proteins.

3-MINUTE SYNTHESIS
One organism's waste product is another's staff of life. Plants and photosynthetic bacteria produce oxygen as a waste product of photosynthesis. Looked at dispassionately, oxygen is toxic and reactive – yet we depend on it as an essential part of our own metabolism, in which we release energy from food. In the microbial world, waste products of metabolism might include such substances as methane, iron and sulphur – yet these will be essential foods for other microbes.

Waste management is a problem for all living organisms. Excretion is the process by which organisms get rid of the wastes generated by metabolism. In general, the process refers to liquid and gaseous wastes – voiding solid waste is known as egestion. Excreted wastes in animals include carbon dioxide, salts and various compounds of nitrogen. Very small organisms, especially if they live in water, excrete from the entire cell or body surface. Larger organisms, such as humans, tend to have specialized organs for the purpose. We excrete carbon dioxide from the lungs and excess salts in sweat from the skin. In small or aquatic organisms, waste nitrogen, generated by the breakdown of proteins and nucleic acids, is released in the form of ammonia, NH_3. This is toxic but also highly soluble in water, so dissolves in the organism's environment before it can accumulate. In many land animals (and also some aquatic ones, such as sharks), nitrogen is excreted through the kidneys as urea, $(NH_2)_2CO$. This gives urine its distinctive odour. Some animals also excrete nitrogen in the form of uric acid. This is insoluble and easily crystallizes out. In humans it can cause problems such as bladder stones and kidney stones, or, if it finds its way to the joints, painful gout.

RELATED TOPICS
See also
MUSCLES
page 64

RESPIRATION
page 94

PHOTOSYNTHESIS
page 98

3-SECOND BIOGRAPHY
WERNER KUHN
1899–1963
Swiss chemist who, with Bart Hargitay, proposed the countercurrent multiplier mechanism for the loop of Henle in the kidney

30-SECOND TEXT
Henry Gee

Human waste is excreted from the lungs, from the skin, and via the kidneys.

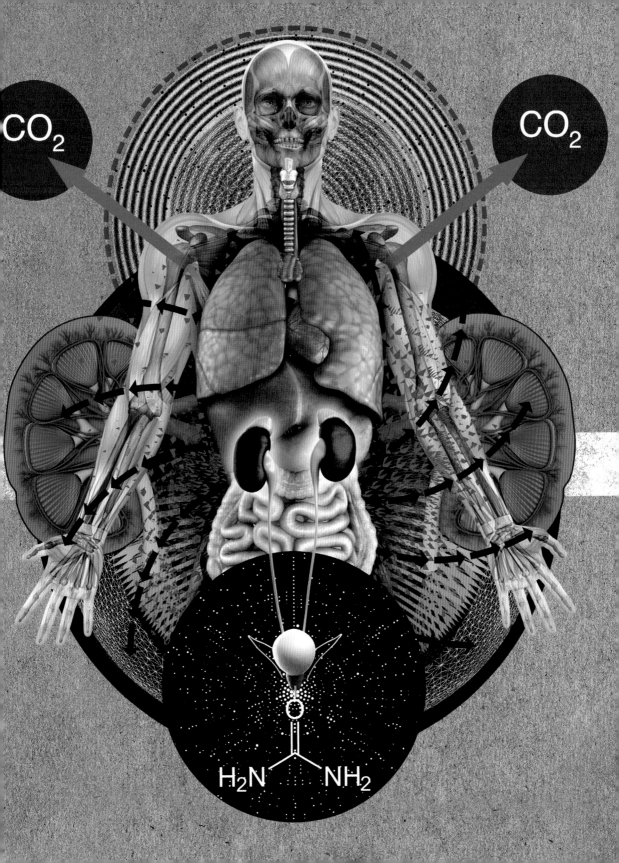

CELLULAR SENESCENCE & DEATH

the 30-second theory

Cell death is as important to

normal health as cell division. In many cases cell death is required to balance cell division to ensure that the overall number of cells stays the same. In other cases cells die through damage, loss of nutrients or as a result of viral infection. In each case the death of the cell helps maintain the overall health of the organism. Every day billions of cells die in a process of cell suicide called apoptosis. In apoptosis cells deliberately and carefully kill themselves – dismantling organelles, cell structures and chromosomes and neatly packaging themselves for safe disposal by specialized cells called macrophages. This occurs when a cell, in particular cell DNA, has become irreparably damaged; rather than continue and risk a malfunction that might lead to a disease such as cancer, the cell kills itself. In some cases cells may be damaged but instead of either repairing the damage or undergoing apoptosis, the cell takes a third path and becomes senescent. Cellular senescence occurs when a cell becomes post-mitotic – in other words it no longer divides and instead enters a period of irreversible growth arrest. Cellular senescence becomes more common as organisms get older and is thought to be a major cause of ageing.

3-SECOND DISSECTION
Billions of cells die every day in multicellular organisms as a normal part of maintaining health and avoiding disease.

3-MINUTE SYNTHESIS
Lymphocytes are immune cells in vertebrates that are capable of recognizing specific microorganisms. To achieve this specificity lymphocytes are produced with random receptors capable of recognizing any potential pathogen. However, this randomness also means that lymphocytes can react to the organism's own cells; they are therefore tested for self-reactivity before release and any that fail this test are killed through apoptosis. More than 90 percent of immune cells die in this way.

RELATED TOPICS
See also
CELLS & CELL DIVISION
page 54

IMMUNITY
page 60

3-SECOND BIOGRAPHIES
JOHN SULSTON
1942–
British biologist who identified cells in nematode worms that undergo programmed cell death (apoptosis) as a normal part of the development of the organism

H. ROBERT HORVITZ
1947–
American biologist who identified key genes which control the process of apoptosis and are common to organisms from flies to humans

30-SECOND TEXT
Phil Dash

In apoptosis a cell breaks down according to a precise programme; it is then disposed of by macrophages.

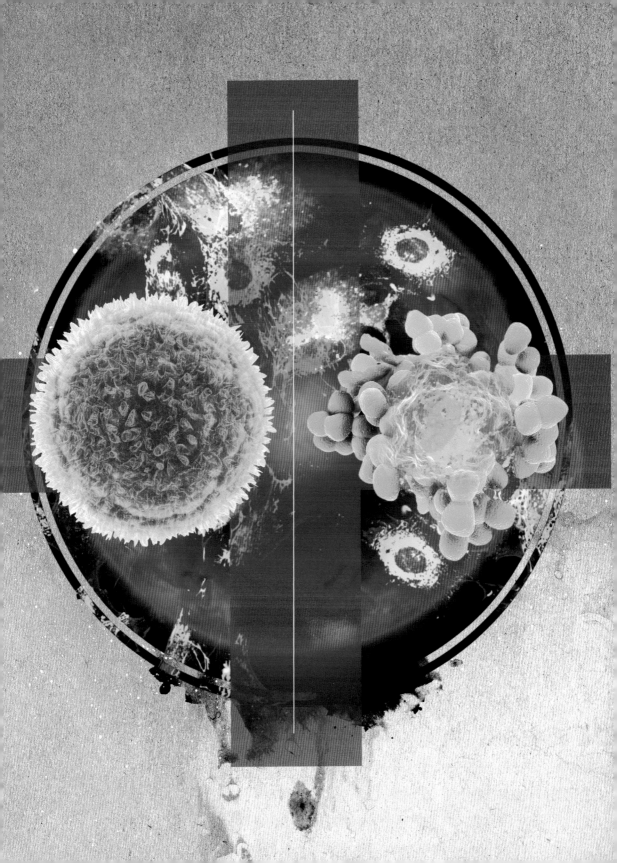

BIOFUELS
the 30-second theory

Biofuels are fuels derived from

biological materials. They are in theory renewable, carbon-neutral and a viable alternative to fossil fuels. Their popularity has increased because of rising oil prices, demands for energy security and the need to reduce greenhouse-gas emissions. Solid biofuels, obtained from fast-growing trees such as poplar and willow or perennial grasses such as switchgrass and miscanthus, are burnt to produce energy. Bioethanol is produced by fermenting sugars and starches in crops like maize, sugar cane and sugar beet. Biodiesel is synthesized from animal fats or vegetable oils. Methane is produced from biodegradable wastes by anaerobic digestion. Unfortunately, biofuel production can be detrimental to the environment, biodiversity and food security. Vast areas are necessary for energy crops to replace fossil fuels completely. The demands of energy crops for irrigation water, mineral fertilizers and agrochemicals can damage the environment, and conversion of savannah or rainforests to biofuel production reduces biodiversity and causes immediate greenhouse-gas emissions that may not be balanced for decades. Using conventional crops as biofuels promotes commercial competition between food and fuel.

3-SECOND DISSECTION
Renewable biofuels save fossil fuels, but growing 'energy crops' can have adverse environmental consequences and takes land from food production.

3-MINUTE SYNTHESIS
Biofuels are renewable alternatives to fossil fuels. However, their production can have adverse consequences for the natural environment, biodiversity and food security. To address these issues scientists are developing so-called 'advanced' biofuels, whose production has less impact on the environment and does not compete with food supplies. Strategies include the production of biofuels from waste products, bioethanol from cellulose and precursors for liquid biofuels from microalgae grown on non-agricultural land.

RELATED TOPICS
See also
CONTROVERSY: GM
ORGANISMS
page 88

METABOLISM
page 100

3-SECOND BIOGRAPHIES
HENRY FORD
1863–1947
American industrialist and founder of the Ford Motor Company, which produced a popular Model T Ford that could run on ethanol

30-SECOND TEXT
Philip J. White

Can we save the world with biofuels? They may reduce the need for fossil fuels, but often occupy land needed for growing food.

EVOLUTION

EVOLUTION
GLOSSARY

archaea In the past archaea were generally called archaebacteria and viewed as a subgrouping of bacteria but they are now known to be distinct. Single-celled prokaryotes (they have no nucleus or other compartments within the cell), archaea have genes and metabolic pathways similar to those found in eukaryotes.

bacteria Group of microscopic organisms, each consisting of just one cell. Bacteria are classified according to whether they need oxygen (aerobic bacteria) or not (anaerobic bacteria). They are also split into groups according to their shape: for example, spiral (*spirillum*), spherical (*coccus*) and rod-like (*bacillus*). Cyanobacteria, also known as blue-green bacteria, derive energy through photosynthesis.

cell Smallest unit of an organism, typically but not always consisting of a nucleus and cytoplasm (the part of a cell that surrounds the nucleus and is enclosed by the cell's outer membrane).

'descent with modification' Charles Darwin's phrase for the way in which, according to their environmental circumstances, species evolve in different directions from common ancestors.

DNA Deoxyribonucleic acid, a molecule that carries the coded genetic information that transmits inherited traits. DNA is found in the cells of all prokaryotes and eukaryotes.

eukaryotes Organisms or cells that have a discrete nucleus, as opposed to prokaryotes (single-celled organisms that do not have a discrete nucleus or other structures or compartments).

fungi Group of multicellular eukaryotes, more closely related to animals than plants. There are around 80,000 known fungi species and these include mildews, yeasts, mushrooms and toadstools.

imprint on Behaviour of a young animal when it makes a connection with the first animal or person it encounters, judging it to be an object worthy of trust.

mutualism and parasitism Mutualism is an interaction between two species that benefits both, while in parasitism one species (the parasite) lives on or inside another (the host), and gets its nutrients from it. Parasites on the human body (ectoparasites) include fleas and lice; parasites in the body (endoparasites) include some bacteria and tapeworm.

natural selection Process through which those organisms that are best adapted to their environment will generally survive and have more offspring. In the theory of English naturalist Charles Darwin, natural selection was one of the key mechanisms – along with genetic drift (random fluctuations in frequency of genetic variation in a population), migration (movement of groups) and mutation (change in a gene's structure) – by which evolution worked.

protists Group of distantly related microscopic organisms, each usually consisting of a single cell. Some, such as algae, contain chloroplasts and are more like plants, while others, such as amoebae, are more like animals. A third subgroup, which includes yeasts, are closer to fungi.

runaway process Theory of sexual selection developed by English statistician and geneticist Ronald A. Fisher. The theory holds that a trait that is initially useful can become sexually selected to a point where it becomes a disadvantage. Sexual selection here refers to females selecting mates with particular features, with the effect that this selected feature becomes widespread in a population because resulting offspring have the trait. The example often given is long and gorgeous tail plumage in a bird: initially females selected males with longer tails because they could fly better, avoid predators and survive longer; through the runaway process longer tails spread through the population and became longer and longer to the point where they potentially became a disadvantage for survival. The rival 'good genes' theory argues that the long tails are selected because they indicate birds with long tails have good genes – they are so healthy they can 'afford' the luxury of extravagant tail feathers.

speciation Development of one or more new species from an existing species. It typically happens when populations of a species become geographically separated – they are 'allopatric'. It may also occur when two populations that are living near to one another and could interbreed do not interbreed due to differences in behaviour, say. These groups are said to be 'sympatric'.

ADAPTATION & SPECIATION

the 30-second theory

3-SECOND DISSECTION
Natural selection happens when changes in the environment meet a species whose members have varying genetic constitutions.

3-MINUTE SYNTHESIS
Natural selection has no memory or foresight, and doesn't always mean 'improvement' or 'progress'. For example, parasitism may mean that an organism becomes simpler instead of more complex, as natural selection adapts the parasite to the habitat provided by its host. Evolution hasn't stopped with modern humans – over the millennia people have adapted to challenges posed by cooking and farming. For example, many humans can now drink cows' milk as adults, a relatively recent evolutionary innovation.

Natural selection is what we call the combined effects on creatures of inherited variation, having lots of offspring, environmental change and the passage of time. Over the generations, these forces will shape organisms to fit their environment, because only those individuals best adapted to their surroundings will live long enough to reproduce and spread their 'favoured' traits to the next generation. This, in essence, is Darwin's theory of evolution by natural selection. Because of adaptation, life evolved from simpler forms to more complex ones. Occasionally, though, it suits organisms to become simpler – for example, birds become flightless and lose their wings. Evolution is all around us today. Antibiotics may kill all of a species of bacteria except those with a trait for resistance. This spreads, and eventually the species becomes resistant. Sometimes groups of animals or plants become separated from the rest of their species by geography or habit. Each group will evolve according to differences in the environment, eventually becoming a different species. If individuals of different groups were to meet, they would no longer be able to interbreed. This is the process of 'speciation'.

RELATED TOPICS
See also
POPULATION GENETICS
page 40

COEVOLUTION
page 118

CHARLES DARWIN
page 120

3-SECOND BIOGRAPHIES
CHARLES DARWIN
1809–82
English naturalist who proposed the theory of evolution by natural selection in his book *On The Origin of Species* in 1859

THEODOSIUS DOBZHANSKY
1900–75
Ukrainian-born American geneticist who was one of those who fused modern genetics with Darwinian selection to create evolutionary biology as we know it today

30-SECOND TEXT
Henry Gee

Darwin's studies included explaining how hummingbirds adapted by developing a long beak that could reach deep into flowers.

SEXUAL SELECTION

the 30-second theory

3-SECOND DISSECTION
Sexual selection is natural
selection applied to the
battle of the sexes as
females try to find the
best potential fathers for
their offspring.

3-MINUTE SYNTHESIS
Why is it that females
end up doing most of the
choosing? It's all a matter
of relative investment.
Males produce millions
of sperm. These are
individually cheap, and
it is in a male's interests
to impregnate as many
females as he can. Females,
though, produce relatively
few eggs. These are
expensive to make, and
because females generally
get burdened with parental
care, it is in their interests
to be choosier than males
about potential mates.

It's important that creatures
choose the best potential mate to maximize
their chances of passing on their genes. This
process of choice allows for what Darwin
realized was a special case of natural selection,
which he called sexual selection. The elaborate
train of the peacock serves no useful function,
he thought, other than to attract the attention
of the less gorgeously attired peahens. The
more attractive the peacock, the more
successful he would be, and he would father
more chicks. The problem, however, is for the
peahen – does beautiful plumage necessarily
translate into successful fatherhood? There are
three main theories to explain this. The 'good
genes' theory holds that attractive display traits
are linked with good health and resistance to
disease. Cockerels infested with parasites, for
example, look duller than their healthier peers.
The 'handicap' principle holds that a male is so
genetically healthy that he can 'afford' to boast
expensive display traits. The 'runaway' process
holds that a male display trait might become
linked, fortuitously, with the female preference
for that trait. The trait itself might initially be
almost random, explaining the vast array of male
display characters seen in nature, from courtship
songs to beautiful plumage to (allegedly) bright
red sports cars.

RELATED TOPICS
See also
ADAPTATION AND SPECIATION
page 114

CHARLES DARWIN
page 120

BEHAVIOUR
page 124

3-SECOND BIOGRAPHIES
RONALD A. FISHER
1890–1962
English statistician and
geneticist who helped put
natural selection on a secure
mathematical footing, and
also pioneered the concept
of the 'runaway' process of
sexual selection

MARLENE ZUK
1956–
American biologist who, with
Bill Hamilton, worked out the
'good genes' hypothesis for
sexual selection

30-SECOND TEXT
Henry Gee

*Male courtship displays
have evolved in
response to female
mate-selection.*

COEVOLUTION

the 30-second theory

Living things do not evolve in isolation. The evolution of all living things affects, and is affected by, the evolution of the organisms that surround them. Sometimes this evolution is antagonistic, like an arms race; at other times, it is for their mutual benefit. Either way, the process is called 'coevolution'. When predators catch the weakest prey, the stronger prey survive to reproduce. In that way, gazelle evolve to be ever more alert to the likelihood of attack by ever-speedier cheetah. Gazelle and cheetah thus 'coevolve'. On a larger coevolutionary scale, flowering plants and insects have coevolved over at least 125 million years to a world-spanning system of mutual benefit. Because plants cannot seek sexual partners, they must rely on other agents to disperse their male sex cells – pollen – to fertilize the female parts of plants. This is why plants have evolved colourful, scented flowers, with lures such as nectar, to attract pollinating insects. In our own history, the domestication of animals such as cattle, sheep and especially dogs can be said to have altered our own evolutionary trajectory. We are different from our ancestors in many ways. We and our animals have therefore coevolved.

3-SECOND DISSECTION
No species is an island. Because creatures are connected in ecological networks, the evolution of any one species will affect that of any other.

3-MINUTE SYNTHESIS
Coevolution covers a multitude of different types of relationship. When a plant species can be pollinated by a variety of insects, each of which can visit several different plant species, coevolution is looser. However, some species have a closer, more dependent relationship in which each relies on another for survival (mutualism); elsewhere, creatures depend on others for survival in such a way that the host is harmed (parasitism). Sometimes the lines between different types of coevolution are hard to draw.

RELATED TOPICS
See also
ADAPTATION & SPECIATION
page 114

MUTUALISMS
page 122

FOOD WEBS
page 138

3-SECOND BIOGRAPHY
LEIGH VAN VALEN
1935–2010
American biologist who likened coevolutionary arms races to the Red Queen's Race in Lewis Carroll's *Alice Through The Looking-Glass*, in which the Red Queen tells Alice that one has to run as fast as possible to stay in the same place

30-SECOND TEXT
Henry Gee

Gazelles evolved to be more alert and cheetahs faster; flowers to be more attractive to pollinating insects. Coevolution puts species' development into context.

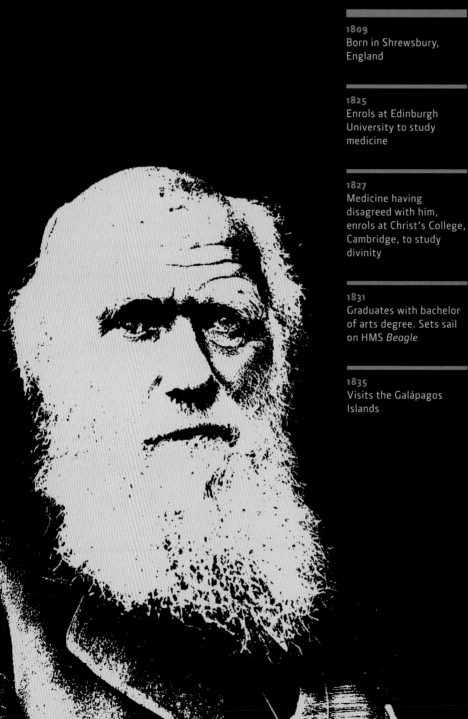

1809
Born in Shrewsbury, England

1825
Enrols at Edinburgh University to study medicine

1827
Medicine having disagreed with him, enrols at Christ's College, Cambridge, to study divinity

1831
Graduates with bachelor of arts degree. Sets sail on HMS *Beagle*

1835
Visits the Galápagos Islands

1836
The *Beagle* returns to England

1839
Publishes *The Voyage of the Beagle*; marries Emma Wedgwood, his first cousin

1842
Settles at Down House in Kent

1858
Darwin's paper on natural selection read at the Linnean Society in London, alongside one by Wallace

1859
Publishes *On the Origin of Species*

1871
Publishes *The Descent of Man*

1882
Dies and is buried in Westminster Abbey

CHARLES DARWIN

Charles Darwin was born in Shrewsbury, England, in 1809, to a doctor, Robert Darwin, cousin of the Wedgwood pottery family. A hearty, outdoorsy type, fond of hunting and shooting, Charles was initially very different from his famous grandfather Erasmus, a philosopher who thought deeply about natural history. Educated at Shrewsbury School, Charles was sent to Edinburgh to study medicine, at which he was a failure. Too squeamish for anatomical demonstrations, he spent much time at lectures by radical thinkers such as Robert Grant, and outdoors studying natural history. His despairing father sent him to Cambridge to study for the last resort of dissolute sons – divinity. It was there Darwin met the Rev. John Stevens Henslow, who was impressed more by Darwin's acute eye as a naturalist than by his devotion to God.

It was Henslow who was responsible for Darwin's lucky break. Henslow recommended Darwin for the position as unpaid companion to Captain Robert FitzRoy on the survey ship HMS *Beagle*, then about to set off for a five-year round-the-world voyage. The *Beagle* set sail in 1831. Darwin's collections of fossils and other natural history specimens gathered during the voyage, especially from South America, were to revolutionize biology. The notes he took on how species differed between the adjacent islands in the Galápagos archipelago germinated in his mind into what he would call 'descent with modification', in which species would evolve in their different directions from common ancestors, according to their environmental circumstances.

After his return in 1836 he wrote *The Voyage of the Beagle*, a bestseller, married his cousin Emma and settled at Down House in Kent to raise a large family and get his thoughts in order. Ever slow to write up his findings, his discovery of what he called 'natural selection' was almost scooped by Alfred Russel Wallace, a young naturalist collecting in the East Indies. Their discoveries were announced together in 1858. The following year Darwin published his masterpiece *On The Origin of Species*, which became a runaway hit and the foundation of modern evolutionary understanding. He published many more books including *The Descent of Man*. He died in 1882 and was buried in Westminster Abbey.

Henry Gee

MUTUALISMS

the 30-second theory

The poet Alfred, Lord Tennyson, described nature as 'red in tooth and claw'. Although it's true that organisms kill, eat, decompose and parasitize one another, they can also form alliances of mutual support. Such associations are called mutualisms, and – it turns out – they are vital to all life. Some mutualisms are close to home. Some of the bacteria that live in our gut could not survive outside it, and pay their rent with all kinds of health benefits we are only beginning to understand. The bacteria in the guts of cattle produce the enzyme cellulase, without which the cows could not break down the tough cell walls of the plants they eat. Most land plants survive thanks to fungi called mycorrhizae, which clothe their roots. The fungi scour the soil for minerals, which they pass to the plants in exchange for nutrients the plants gain from photosynthesis. Many plants rely for their pollination on certain insects, which in turn may be nourished by the plant, and even, in cases such as the fig wasp and the fig, housed by it in specially modified structures.

3-SECOND DISSECTION
Mutualisms are ubiquitous associations between organisms that cooperate for mutual benefit.

3-MINUTE SYNTHESIS
The cells in our bodies started more than 2 billion years ago as mutualisms between prokaryotes (bacteria and archaea). Mitochondria, the energy factories of cells, are related to alpha-proteobacteria. Chloroplasts, the structures in plant cells in which photosynthesis takes place, were once cyanobacteria (blue-green algae). The nucleus probably started out as an archaeon. Today, the cell is an indivisible unit and no component could survive on its own independently.

RELATED TOPICS
See also
BACTERIA
page 18

FUNGI
page 24

COEVOLUTION
page 118

3-SECOND BIOGRAPHY
LYNN MARGULIS
1938–2011
American biologist who proposed the 'symbiogenesis' theory in which eukaryotic cells began as mutualisms between different prokaryotes

30-SECOND TEXT
Henry Gee

Cooperation is widely found in nature, from the fig and the fig wasp to buffalos and the cattle egrets that follow them.

BEHAVIOUR

the 30-second theory

Living things do not sit around doing nothing. They constantly monitor their surroundings and react to them. In other words, they behave. Bacteria respond to light and the presence of chemicals by moving towards or away from the stimulus. Plants detect gravity, chemicals, even the presence of other plants – and respond. The word 'behaviour', however, is generally associated with animals, especially those such as insects (ants, bees and so on); molluscs (such as octopuses); and vertebrates (such as ourselves), all of which have brains and nervous systems. For such animals the environment is rich in sensation, signalling either opportunity or threat. Most animals behave in stereotypical ways. That is, they always respond to certain stimuli in pre-programmed ways. Such behaviour represents an adaptive response to specific situations. All gazelles will run away from a cheetah, because those that run towards cheetahs, ignore their threat or run too slowly will be eaten, and fail to pass their genes to the next generation. Many animals can learn to adopt new behaviours in certain situations. But only a few, including humans, some apes, dolphins, elephants and crows, are able to reflect on their behaviour, an attribute associated with a sense of 'self'.

3-SECOND DISSECTION
Behaviour is a term for an organism's range of responses to its environment.

3-MINUTE SYNTHESIS
A recent trend in the study of animal behaviour is to investigate how and why animals vary in their 'personality' – which need not imply a sense of 'self'. That is, some animals can be brave and outgoing, whereas others in the same species tend to be shy and secretive. Although personality as a 'syndrome' of behaviour can be modified by learning and experience, there is also likely to be a strong genetic component.

RELATED TOPICS
See also
ADAPTATION & SPECIATION
page 114

SEXUAL SELECTION
page 116

JANE GOODALL
page 142

3-SECOND BIOGRAPHIES
KONRAD LORENZ
1903–89
Austrian zoologist who pioneered the study of animal behaviour, particularly how some newborn animals 'imprint' on the first moving object they see

JANE GOODALL
1934–
English zoologist whose 55-year study of chimpanzees in Tanzania is the basis of much we know about the behaviour of our closest animal relatives

30-SECOND TEXT
Henry Gee

Chimpanzees in different areas have different ways of behaving. This is seen in their tool use, grooming and feeding.

GLOBAL PHYLOGENY

the 30-second theory

The only illustration in Charles Darwin's seminal work *On the Origin of Species* was a family tree – not of people, but of species. An evolutionary family tree is called a phylogeny. As individual species split into two and continue to evolve, it's easy to see that what starts as small twigs becomes branches and trunks of a global phylogeny, in which every living thing is related to every other. Penguins to people, bacteria to beech trees – we are all cousins. Nobody knows when or where life first emerged, but the biochemical similarities between living cells, especially the genetic material DNA, shows that all life alive today had a common ancestor. The simplest cells, related to modern bacteria, appeared more than 3.5 billion years ago. Today, life is split into two major domains – the bacteria on one side, the archaea on the other. The eukaryotes – including animals, plants, fungi and protists – stem from within the archaeal tree. Since the 1970s, advances in theory as well as computer and lab technology mean that phylogeny is based on molecular similarity rather than gross anatomical resemblance. Now, molecular trawls of the Earth's environment, from the deep ocean to garden ponds, are uncovering DNA sequences of hitherto unknown kinds of life that contribute to the ever-growing tree of life.

3-SECOND DISSECTION
A phylogeny is an evolutionary family tree. The global phylogeny seeks to unite all life into a single, tree-like diagram.

3-MINUTE SYNTHESIS
Until the 1970s, trying to understand evolutionary relationships (phylogenetic reconstruction), was a haphazard business. Competing phylogenies were hard to assess objectively. Then a method called 'phylogenetic systematics' or 'cladistics' put phylogenetic reconstruction on a much firmer scientific footing. The ability to sequence DNA from organisms, starting at about the same time, unlocked a vast resource of evolutionary information on which to base ever more secure phylogenies.

RELATED TOPICS
See also
ORIGINS OF LIFE – VIRUSES
page 14

ADAPTATION & SPECIATION
page 114

CHARLES DARWIN
page 120

3-SECOND BIOGRAPHIES
WILLI HENNIG
1913–76
German entomologist who invented 'phylogenetic systematics' or 'cladistics', a more objective method of estimating evolutionary relationships

EMILE ZUCKERKANDL
1922–2013
French biologist who, with Linus Pauling, founded the discipline of molecular phylogeny, using molecular information to calculate evolutionary relationships

30-SECOND TEXT
Henry Gee

The archaeal family tree plots the development of all species that are not bacteria – both plants and animals.

WHY WE AGE

the 30-second theory

Why do we grow old and die?

Scientists are beginning to tackle the question, but they are very far from agreement. There are at least three answers. One holds that because resources are finite, individual creatures can devote energy to one activity only at the expense of another: there is evidence of a trade-off between reproduction and longevity. That is, creatures that reproduce early in life, and have lots of offspring, tend to senesce (show signs of ageing) and die younger than those that reproduce later on and have fewer offspring. This is why mice live for a year or two, but elephants and people can live for many decades. Related to this is the second possible answer: genes favourable in youth become deleterious when we age. The third view is summarized as the 'live fast, die young' hypothesis, which holds that ageing is related to speed of metabolism. Biochemical activity in the body generates toxic by-products, which if left unchecked can damage the genetic material DNA. These so-called 'reactive oxygen species' are mopped up with chemicals called antioxidants, such as vitamin C. Connected to this metabolism idea is the view that lifespan is related to diet. Severe dieting has been shown to improve lifespan in animals such as roundworms. All three answers explain different aspects of why we age.

3-SECOND DISSECTION
Mortality has preoccupied people since the dawn of history. It is the theme of the Gilgamesh epic told more than 4,000 years ago, and of much philosophy and religion since.

3-MINUTE SYNTHESIS
An intriguing question is why creatures age at different rates. Your hamster will die within two years, when humans are still infants; your cat or dog within 20, when humans are barely fully mature. However, there is no simple relationship between diet, metabolism, reproduction and ageing. For example, nobody knows why birds tend to live longer than mammals of similar mass.

RELATED TOPICS
See also
CELLS & CELL DIVISION
page 54

CANCER
page 84

CELLULAR SENESCENCE
& DEATH
page 106

3-SECOND BIOGRAPHIES
LEONARD HAYFLICK
1928–
American biologist who discovered that animal cells can divide a finite number of times, known as the 'Hayflick limit'

CYNTHIA KENYON
1955–
American biologist who has pioneered studies on the molecular biology of ageing, using as a model the laboratory roundworm *Caenorhabditis elegans*

30-SECOND TEXT
Henry Gee

Want to live forever? In the longevity race no animal can keep up with the tortoise.

ECOLOGY

anthropogenic climate change Major, long-term change in the Earth's weather patterns and average temperatures seen from the mid/late twentieth century onwards and attributed by climatologists to higher levels of carbon dioxide in the atmosphere resulting from the burning of fossil fuels (coal, gas, oil, etc). Effects have been seen, for example, in retreating glaciers, shrinking ice sheets in the Antarctic and Greenland, rising sea levels and changing rainfall patterns – including an increase in extreme occurrences of heavy rainfall over North America.

autotrophs Organisms that make the organic materials they need for nutrition from inorganic substances such as carbon dioxide and nitrates. Green plants are autotrophs. In contrast, heterotrophs need organic substances – typically plants or animal flesh – to meet their nutritional needs. Humans are heterotrophs.

biodiversity Variety of plants, animals and microorganisms in a habitat. In general usage, biodiversity means the variety of species in a particular habitat or on Earth.

biomass Mass of living organisms of a particular type or in a given area/volume. In discussions of energy needs, the term is also used to refer to material to be used as fuel that is derived from living (or recently dead) organisms.

biotic homogenization The way localized species extinction and the movement of species between habitats leads to diverse locations becoming more similar in terms of diversity of species.

bovine tuberculosis Form of tuberculosis (a bacterial disease in which tubercles or nodules grow in the lungs and other tissues) that affects cattle. It is caused by the spread of the aerobic bacterium *Mycobacterium bovis*. The disease can affect other mammals including badgers, deer, pigs and humans; some authorities contend that badgers are major culprits in spreading the disease to cattle and advocate culling badgers, but there is controversy over whether this is effective.

demographics Statistics of birth and death rates of populations and groups within them.

desertification Process by which 'dryland' ecosystems of Earth become more arid, losing vegetation, wildlife and areas of water. Drylands are ecosystems that are very low in or lack water – they include deserts, but also savannahs, grasslands, scrublands and similar regions. Causes of desertification include climate change, deforestation and overgrazing by animal flocks.

ecosystem Biological community of organisms, its associated environment and the interactions between the organisms.

endangered species A species officially categorized as likely to become extinct. Endangered species are listed on the International Union for Conservation of Nature (IUCN) Red List. A species becomes extinct when it has no living members.

genomics Study of an organism's genome (genetic material), focusing on its evolution, function and structure.

inbreeding Breeding among people or between animals that are closely genetically related. In small or isolated populations inbreeding can lead to inbreeding depression: the population suffers a decline in biological fitness.

natural selection The process through which those organisms that are best adapted to their environment will generally survive and have more offspring, part of the theory of evolution proposed by English naturalist Charles Darwin.

niche In ecology, the role and status (position) of a species within an ecosystem. It refers to all the species' interactions with others in the ecosystem – for example, competing for food, acting as a predator, living as a parasite. Two species cannot achieve stable coexistence if they attempt to occupy the same niche.

palaeontologist A scientist who studies extinct organisms through fossils.

primatologist A zoologist (a scientist who studies animals) who specializes in studying primates, the order of mammals that includes monkeys, apes and humans.

rewilding Conservationist strategy that sets out to protect natural processes and restore areas of wilderness. A key aspect of the strategy is to reintroduce large predator species to wilderness areas. Proponents of the idea argue in favour of reintroducing species such as wolves and lynx to 'rewilded' areas of, for example, the UK or North America.

species Set of living organisms whose members can interbreed and produce fertile offspring. Species is the eighth category in the scientific classification system.

BIOGEOGRAPHY

the 30-second theory

3-SECOND DISSECTION
Little is random about the distribution and abundance of species across the globe; each has been affected by climate and geography, but ultimately, time is the most important factor.

3-MINUTE SYNTHESIS
Until recently species distributions were the outcome of changes that occurred at geological timescales. Today, whether accidentally or deliberately, we are introducing species to regions where they would not normally have spread. Burmese pythons in the Florida Everglades, Asian parakeets in London's suburbs and South American cane toads in Australia are visible examples of how in a few short centuries we have reshaped ecosystems – with consequences we are only just beginning to understand.

Species are not spread evenly around the globe, but there are clear patterns in their distributions that suggest similar processes affect their presence or absence. The nearer we are to the Tropics, the more species we find, a result of better conditions for plant growth and more stable climates. There are fewer species as altitude increases, for similar reasons. Both isolation and island size affect diversity, with distant, small islands having fewer species than those closer to the mainland. Barriers to movement, and isolation over extended periods of time, allow the evolution of new species, which is why places such as Australia, Hawaii and Madagascar are so important in terms of global biodiversity. Each of these patterns is easy to explain, but harder to understand are some of the peculiarities of species distributions. In Africa we have the ostrich, in South America the rhea, Madagascar the now extinct elephant bird and in Australia we find the emu. These are all similar, closely related species. Just how did they end up so far apart? Modern theory suggests that their common ancestor inhabited Gondwanaland, the super-continent that broke apart perhaps 180 million years ago in Cretaceous times. As the land masses moved to their current positions due to continental drift, they carried the descendants along with them.

RELATED TOPICS
See also
FOOD WEBS
page 138

CLIMATE CHANGE BIOLOGY
page 144

EXTINCTION
page 148

3-SECOND BIOGRAPHIES
GEORGES-LOUIS LECLERC, COMTE DE BUFFON
1707–88
French naturalist who noted how climate affected species' distributions

ALFRED LOTHAR WEGENER
1880–1930
German geophysicist who noted how continents fitted together like a jigsaw, and postulated that they had once been linked

30-SECOND TEXT
Mark Fellowes

Ancestors of many of today's similar but far-flung species lived on a united 'supercontinent'.

POPULATION ECOLOGY

the 30-second theory

3-SECOND DISSECTION
Population ecology aims to understand why some species are plentiful, while others face extinction.

3-MINUTE SYNTHESIS
For conservation biologists, estimating the minimum viable population of a rare species is crucial for understanding its likelihood of extinction. This estimate takes into account the effects of random chance, major environmental catastrophes and inbreeding, which are more likely to affect small populations. It has been calculated that around 4,000 individuals is the minimum needed in a population to ensure a 95 per cent chance of survival for 100 years. Many endangered species have much smaller populations than this.

How many people will inhabit the Earth in 2050? Predicting this is relatively straightforward (the United Nations says there will be around 9.7 billion people), because we know the demographics of our human populations. We record births and deaths, when, why and where they occur. This is not so for other species. Do badgers affect rates of bovine tuberculosis in cattle in the UK? Are we unsustainably harvesting sea fish? Will wild tigers be extinct in 20 years? Answering each of these questions relies on having demographic information that can be used in mathematical and statistical models that predict changes in a species' population size. These models show how such changes are affected by factors such as the environment, competition for resources, predation and diseases. The simplest population growth model is one of exponential growth, where individuals have offspring, who in turn breed, adding many more individuals over time. But populations cannot continue to grow without end; food becomes limiting, so population size is determined by the carrying capacity of the environment. When a population is above this it will fall; if below, it will grow towards it. Predators or environmental change can introduce variation, leading to cyclical rises and falls in population size.

RELATED TOPICS
See also
MUTUALISMS
page 122

FOOD WEBS
page 138

3-SECOND BIOGRAPHIES
THOMAS ROBERT MALTHUS
1766–1834
English clergyman who noted that populations are limited by resources, greatly affecting Charles Darwin's thoughts on natural selection

GEORGII FRANTSEVICH GAUSE
1910–86
Russian ecologist who suggested that no two species can inhabit the same niche, as one will always drive the other to extinction

30-SECOND TEXT
Mark Fellowes

Conservationists report that only a few thousand tigers survive in the wild – will they go the way of the dodo?

FOOD WEBS

the 30-second theory

3-SECOND DISSECTION
The world is a dangerous place, where every species is predator or prey, and for those in the middle of a food chain, both.

3-MINUTE SYNTHESIS
Elton provided modern ecology with many founding principles. Most important was the pyramid of numbers: the biomass of plants (primary producers) is much greater than that of herbivores (primary consumers); in turn, there are fewer predators (secondary consumers) and predators of predators. This reflects the inefficiency of energy conversion as prey are consumed and turned into new predators. Big predators at very high trophic levels are rare.

No species exists in isolation; each is part of a complex network of interactions. Some of these interactions are positive, where the species benefit, but most are negative, with other individuals or species competing for resources or acting as natural enemies. Capturing how assemblages of organisms interact is the emphasis of community ecology, and its most simple tool is the food web. The first food web was drawn by Charles Elton in 1923. While a student at Oxford University, Elton took part in an expedition to Bear Island, near Spitsbergen. Along with a botanist, V.S. Summerhayes, Elton studied who ate whom in this tundra environment. This laid the groundwork for Elton's key insight: that you could see how energy flowed from plant to herbivore to predator in simple food chains, and that these chains intermingled in a food web. Food webs have been much improved since Elton's first description. The most fundamental change has been to place species that have similar ways of gaining energy in what are known as trophic layers. Here, plants, which make their own energy through photosynthesis, are at the bottom, linked to their herbivores on the next layer, which in turn are connected to predators above.

RELATED TOPICS
See also
MUTUALISMS
page 122

POPULATION ECOLOGY
page 136

ECOSYSTEM ENERGETICS
page 140

3-SECOND BIOGRAPHIES
CHARLES SUTHERLAND ELTON
1900–91
British zoologist who helped transform ecology into a quantitative science

RAYMOND LAUREL LINDEMAN
1915–42
American ecologist who suggested that only around 10 per cent of energy in one trophic level gets transferred up to the next one

30-SECOND TEXT
Mark Fellowes

Energy flows up the pyramid, from one trophic layer to the next – plants to herbivores to predators.

ECOSYSTEM ENERGETICS

the 30-second theory

RELATED TOPICS

See also
FOOD WEBS
page 138

EXTINCTION
page 148

30-SECOND TEXT
Mark Fellowes

3-SECOND DISSECTION
Plants transform enormous quantities of energy into accessible organic material, and humans consume the greatest proportion of it.

3-MINUTE SYNTHESIS
Patterns of NPP vary across the globe, with wetter, warmer regions having much higher rates than desert or arctic regions. This explains a great amount of the variation in patterns of biodiversity across the globe. Terrestrial regions with higher NPP have more species, explaining why tropical rainforests are such biodiversity hotspots. Oddly, in marine environments there is an inverse relationship between NPP and biodiversity due to ocean current mixing. This boosts productivity but lowers diversity.

Almost all of the energy that powers life on Earth comes from the sun, converted by photosynthesis into organic compounds. The total rate of energy converted by the autotrophs (species that make their own energy) is the Earth's Gross Primary Productivity, which once we remove the energy expended to allow this to happen, gives us Net Primary Productivity (NPP), the rate at which energy is transformed into organic matter. This productivity accumulates over time, as biomass, and is a measure of the energy that heterotrophs (most non-photosynthesizing species) can exploit. Humans make up around 0.5 per cent of heterotroph biomass, yet by some estimates we use more than 23 per cent of global NPP. It is estimated that South-east Asia and Western Europe devour more than 70 per cent of their regional NPP. We directly consume plant biomass for food, for fuel and as materials for manufactured goods. We further reduce NPP by converting forests, which are highly efficient, into less efficient croplands, by consuming herbivores rather than plants, and by degrading habitats through desertification, urbanization and pollution. As our population grows and our standards of living rise, these demands on our planet's ability to feed us will only become harder to meet.

Clearing forests to provide grazing land for herds reduces the land's energy efficiency.

1934
Born Valerie Jane Morris-Goodall to Mortimer Morris-Goodall and Vanne Joseph in London

1952
Leaves school to work as a secretary

1958
Studies primate biology in London

1960
Begins studies of chimpanzees in Gombe Stream National Park, Tanzania (then Tanganyika)

1962–5
Completes PhD in ethology at Newnham College, Cambridge

1964
Marries Dutch wildlife photographer Baron Hugo van Lawick

1974
Divorced from van Lawick

1975
Marries Derek Bryceson, director of Tanzania's national parks

1977
Sets up Jane Goodall Institute

1980
Bryceson dies of cancer

1991
Starts 'Roots and Shoots' youth education programme in Virginia, USA

1996
Awarded the Zoological Society of London Silver Medal

2004
Becomes Dame Commander of the Order of the British Empire

2006
Receives UNESCO 60th Anniversary Medal and membership of the French Légion d'Honneur

JANE GOODALL

One of a trio of primatologists

sponsored by palaeoanthropologists Louis and Mary Leakey – the others being Dian Fossey and Biruté Galdikas – Jane Goodall has achieved widespread public recognition for her work with chimpanzees.

In her early twenties, Goodall travelled to Kenya for a visit to the family farm of her friend Clo Mange. While staying there, Goodall introduced herself to Louis Leakey, who initially engaged her as an assistant, soon proposing that she should embark on research into chimpanzees. Leakey felt a study of primates would help clarify the joint ancestry of great apes and early humans. After a brief period of introduction to the subject in London, Goodall was set to work in Gombe Stream National Park in Tanzania. Two years later, Leakey arranged funding for Goodall to undertake a PhD in ethology – the study of animal behaviour – at the University of Cambridge, despite Goodall's lack of a first degree. Her thesis 'Behaviour of the Free-Ranging Chimpanzee' was completed in 1965.

This began a decades-long study of the social behaviour of chimpanzees at Gombe Stream. Academic opinion is divided on whether Goodall's lack of scientific training was help or hindrance. Her hands-on involvement with the chimpanzees, effectively becoming a member of a chimpanzee troop for a period, reduced the objectivity of her work and may have made changes to the behaviour of the animals.

Goodall's critics also accused her of anthropomorphism, but her enthusiasm and dedication enabled her to make observations of chimpanzee personality and behaviour that could have escaped a more regimented approach.

Goodall's studies brought out previously unknown aspects of chimpanzee life. As well as seeing clear, individual personality traits in the animals, she observed basic tool use – notably 'fishing' for termites using long grass stalks. Similarly, the assumption, fostered in part by zoos, had been that chimpanzees were relatively docile and vegetarian; Goodall uncovered regular consumption of monkeys by the chimpanzees and dramatic levels of violence to maintain troop hierarchies.

In 1986, Goodall attended a conference in Chicago where the extent of chimpanzee habitat reduction was made clear for the first time. Shortly after, she switched her priorities to conservation, organization and publicity, notably in her work with the Jane Goodall Institute, which both arranges studies and helps maintain the chimpanzees' environment. This led to the establishment of a centre for primate studies at the University of Minnesota in 1995. Goodall continues to work tirelessly to promote chimpanzee conservation.

Brian Clegg

CLIMATE CHANGE BIOLOGY

the 30-second theory

The Earth's climate has always varied. There have been times when CO_2 was ten times its present concentration, sea levels were many metres higher and the planet was so warm that tropical plants grew near the North Pole. Current concern over climate change arises because it is rapid and caused by humans. We need to measure the impact on plants and animals of changes in temperature and CO_2 concentration, melting polar ice caps, rising sea levels, decreasing ocean acidity, drought and flooding. Organisms are highly adapted to their environments; with climate change, some contract their range and may even go extinct, while others expand and become problematic. Amphibians – frogs, newts, salamanders – seem particularly sensitive. Extinctions have been associated with habitat loss, disease and climate change. The desert locust, on the other hand, is likely to become more destructive with the predicted increase in extreme precipitation events in North Africa's Sahel region. The timing of biological events in the year is also changing; where different organisms– like flowering plants and pollinating insects – schedule their seasonal activities to coincide, varying responses to climate change may be very disruptive. Biologists must understand climate change responses to protect the Earth's biodiversity.

3-SECOND DISSECTION
Rapid anthropogenic (human-caused) climate change is challenging species' ability to remain closely adapted to their environment – can they evolve quickly enough to keep up?

3-MINUTE SYNTHESIS
The seasons alter:
hoary-headed frosts
Fall on the fresh lap
of the crimson rose;
And on old Hiems'
thin and icy crown
An odorous chaplet of
sweet summer buds
Is, as in mockery, set.
The spring, the summer,
The childing autumn,
angry winter, change
Their wonted liveries;
and the mazed world,
By their increase, now
knows not which is
which.
ACT II, SCENE I
A MIDSUMMER
NIGHT'S DREAM
WILLIAM SHAKESPEARE

RELATED TOPICS
See also
BIOGEOGRAPHY
page 134

ECOSYSTEM ENERGETICS
page 140

EXTINCTION
page 148

CONTROVERSY: THE
ANTHROPOCENE
page 150

3-SECOND BIOGRAPHIES
CAMILLE PARMESAN
1961–
US pioneer in the assessment of the impacts of climate change on wildlife

SIR BRIAN HOSKINS
1945–
British mathematician who championed the subject of meteorology and emphasized the importance of climate change for society

30-SECOND TEXT
Nick Battey

Climate change is altering the environment for species across the globe.

INVASIVE SPECIES

the 30-second theory

Invasive species are those that
colonize new territories and become
problematic: from the flu virus that annually
spreads around the globe, to the harlequin
ladybird currently outcompeting native UK
ladybirds; from the Burmese python that is
damaging biodiversity in the Florida Everglades,
to the greatest invader of all – the human.
Often species that become invasive have been
introduced deliberately in a well-intentioned
attempt to tackle a problem. An example is
kudzu (*Pueraria lobata*), a trailing vine native
to east Asia that was widely planted in the
southern USA to control soil erosion, but which
has since become a pernicious weed there.
But most generally, the globalization of human
movement and activity is associated with
biological invasions. Ballast water carried by
ships is blamed for transporting the zebra
mussel, which originated in the Black Sea but
now infests waterways and lakes in North
America. Imports of ash were probably to blame
for an outbreak of ash dieback in the UK. The
brown tree snake *Boiga irregularis* found its way
in the hold of a ship or plane to the Pacific island
of Guam, where it decimated native bird
populations. Dispersal is facilitated by humans;
establishment and spread, the next phases of
invasion, depend on the traits of the invasive
organism and the ecosystem under threat.

3-SECOND DISSECTION
Increased human
movement around
the planet increases
the potential for
species invasions, so
understanding what turns
an introduction into an
invasion is a hot topic.

3-MINUTE SYNTHESIS
What determines the
success of an invader?
Its own traits, such as
fecundity and dispersal,
are important. Ecosystem
diversity may also play a
role, but resource
availability is likely to be
a key factor influencing
vulnerability to invasion.
Whatever the causes, the
consequence is biotic
homogenization (the way
different locations are
made more similar in terms
of diversity of species): as
with dominant global retail
brands, everywhere on
Earth may one day have
the same few, standardized
plants and animals.

RELATED TOPICS
See also
BIOGEOGRAPHY
page 134

POPULATION ECOLOGY
page 136

CLIMATE CHANGE BIOLOGY
page 144

3-SECOND BIOGRAPHIES
CHARLES ELTON
1900–91
British zoologist and ecologist
whose *The Ecology of
Invasions by Animals and
Plants* (1958) defined the field
of invasion biology

MARK WILLIAMSON
1928–
British biologist and author of
Biological Invasions (1996)

30-SECOND TEXT
Nick Battey

*Humans have invaded
almost every corner of
the Earth, encouraging
the spread of invasive
species.*

EXTINCTION

the 30-second theory

Nobody knows how many

species there are on our planet, but the best estimate is around nine million. Until humans began to dominate, the background biodiversification rate (the rate at which new species arise minus the rate at which they become extinct) had been steady or perhaps increasing slightly. But the extinction rate is now about 1,000 times what it was before humans appeared. We are losing an estimated 11,000–58,000 animal species annually. This sudden loss of species is unusual in the history of the Earth: there have been five previous mass extinctions – 443, 359, 251, 200 and 65 million years ago – with a variety of natural causes; an asteroid impact in the Yucatán peninsula of Central America led to massive global cooling and triggered the last one. The present species extinction crisis, the sixth in Earth history, is caused by human activity. Habitat loss, overexploitation, invasive species and climate change are all associated with the human population explosion. As a result, half of all remaining plant and animal species will be extinct by the end of the 21st century. This will create instability in ecosystems, with highly unpredictable consequences. Nature will only survive the human onslaught if we manage our impacts more effectively than in the recent past.

RELATED TOPICS
See also
CLIMATE CHANGE BIOLOGY
page 144

INVASIVE SPECIES
page 146

CONTROVERSY: THE ANTHROPOCENE
page 150

3-SECOND BIOGRAPHIES
E.O. WILSON
1929–
US champion of biodiversity conservation

A.D. BARNOSKY
1952–
US biologist who highlighted the present mass extinction event

30-SECOND TEXT
Nick Battey

3-SECOND DISSECTION
Extinction is a normal part of evolution but the present rate is probably unprecedented in the history of our planet.

3-MINUTE SYNTHESIS
De-extinction – bringing back extinct species using preserved DNA or sperm – has been proposed for iconic animals such as the woolly mammoth. This may be possible, but it seems an extraordinary way to expend resources when current species are being lost at an alarming rate due to human carelessness.

Humans are responsibl for the sixth mass extinction event in the history of life on Earth.

THE ANTHROPOCENE
the 30-second theory

The Anthropocene is a term that
captures the dramatic increase in the influence of humans on the planet over the past 200 years. The human population has exploded from 1 billion to 7 billion; the discovery and exploitation of fossil fuels have led to a 40-fold increase in energy use and greenhouse-gas emissions have sky-rocketed; land is increasingly dominated by humans; rivers are dammed, the oceans are acidifying and the destruction of species due to human activity will cause the sixth major extinction event in the history of the planet. The scale of human impact is increasing every year, and there are also qualitative shifts: genomics technologies will transform the impact of humans on the living world. All this means that humans have become a global force, on the same scale as volcanic eruptions, asteroid impacts and earthquakes: we determine the fate of our planet as never before. The Holocene, which began at the end of the last glaciation, is at an end; the Anthropocene, which will be witnessed in the geological record millions of years hence by the chemical signals of plastics, mass extinctions, disappearance of forests and increasing sea levels, is upon us. How we learn to cope with the congestion, the cars and the conflict will be signalled in that other distinctive feature of humans: cumulative culture.

3-SECOND DISSECTION
The Anthropocene is proposed as a new geological epoch that reflects the present phase of human domination in the history of our planet.

3-MINUTE SYNTHESIS
The idea of the Anthropocene is controversial because of the emphasis it gives to humans, whose 0.2 million years of existence as modern *Homo sapiens sapiens* is as the blinking of an eye in comparison to the 4.5-billion-year history of our planet. Even working out where it should start is tricky: it could be with the origin of agriculture, the Industrial Revolution or the nuclear era.

RELATED TOPICS
See also
CLIMATE CHANGE BIOLOGY
page 144

INVASIVE SPECIES
page 146

EXTINCTION
page 148

3-SECOND BIOGRAPHIES
PAUL CRUTZEN
1933–
Dutch chemist who championed the use of the term Anthropocene

EUGENE STOERMER
1934–2012
US biologist who invented the term Anthropocene in order to emphasize the impact of humans on the planet

30-SECOND TEXT
Nick Battey

More and more humans, fewer and fewer species of plants and other animals… Is time running out?

APPENDICES

RESOURCES

BOOKS

Campbell Biology
Jane B. Reece et al
(Benjamin Cummings; 9th edition, 2011)

Biology
Peter H. Raven et al
(McGraw-Hill; 7th edition, 2005)

Life: The Science of Biology
David E. Sadava et al
(W.H. Freeman; 10th Edition, 2012)

Biology of Plants
Peter H. Raven et al
(W.H. Freeman, 2004)

Plant Biology
Alison M. Smith et al
(Garland Science, 2010)

The Diversity of Life
Edward O. Wilson
(Harvard University Press, 1992)

The Selfish Gene
Richard Dawkins
(Oxford University Press, 1978)

*Guns, Germs, and Steel: The Fates
of Human Societies*
Jared M. Diamond
(W.W. Norton & Company, 1999)

*Nature's Nether Regions: What the Sex Lives of
Bugs, Birds and Beasts Tell Us About Evolution,
Biodiversity and Ourselves*
Menno Schilthuizen
(Penguin Books, 2015)

The Sixth Extinction: An Unnatural History
Elizabeth Kolbert
(Picador, 2015)

*The Variety of Life: A Survey and a Celebration
of all the Creatures that Have Ever Lived*
Colin Tudge
(Oxford University Press, 2002)

Genetics
Hugh Fletcher et al
(Garland Science; 4th Edition, 2012)

Molecular Biology of the Cell
Bruce Alberts et al
(Garland Science; 6th Edition, 2014)

Developmental Biology
Scott F. Gilbert
(Sinaeur Associates; 9th Edition, 2010)

Biochemistry
Reginald H. Garrett & Charles M. Grisham
(Brobks Cole; 5th Edition, 2014)

Molecular Cell Biology
Harvey Lodish et al
(W.H. Freeman; 7th Edition, 2012)

Microbiology
Simon Baker et al
(Taylor & Francis; 4th Edition, 2011)

WEBSITES

http://www.britishecologicalsociety.org/100papers/100InfluentialPapers.html#2
Overviews the world of ecological research by highlighting 100 papers that have influenced thinking.

Animal Diversity Web:
http://animaldiversity.org/
Insights into animals from a taxonomic perspective.

Royal Society of Biology:
http://www.rsb.org.uk
General information about biology, including careers for biologists.

Encyclopedia.com Biology page
http://www.encyclopedia.com/topic/biology.aspx
Details, references and resources on biology.

iBiology http://www.ibiology.org
The wider picture of biology, including insights from top biologists and educational information.

Biology Reference
http://www.biologyreference.com
A world of facts about the world of life.

EDITORS

Nick Battey is Professor of Plant Development at the University of Reading. He has published extensively on pure and applied plant biology and has a strong interest in the history of biology. He is co-author of the book *Biological Diversity: Exploiters and Exploited*. Nick completed a BSc in Plant Science at the University of Wales and a PhD in Plant Developmental Biology at Edinburgh University.

Mark Fellowes is Professor of Ecology at the University of Reading. His broad research interests range from how insects evolve resistance to their natural enemies through to the consequences of urbanization on the abundance and diversity of wildlife. He was lead editor of the book *Insect Evolutionary Ecology*. Mark completed his BSc in Zoology and his PhD in Evolutionary Biology at Imperial College London before moving to Reading, where he is currently Head of the School of Biological Sciences.

CONTRIBUTORS

Brian Clegg read Natural Sciences, focusing on experimental physics, at the University of Cambridge. After developing hi-tech solutions for British Airways and working with creativity guru Edward de Bono, he formed a creative consultancy advising clients ranging from the BBC to the Met Office. He has written for *Nature*, *The Times*, and the *Wall Street Journal* and his published titles include *A Brief History of Infinity* and *How to Build a Time Machine*.

Phil Dash is Associate Professor in Cell Biology at the University of Reading. His research interests include excessive cell movement in cancer. He obtained his BSc in Zoology from the University of Reading, and an MSc in Toxicology and a PhD in Cancer Studies from the University of Birmingham.

Henry Gee is a senior editor at the science journal *Nature*. He has published widely within the biological sciences, and specializes in evolutionary issues, including *The Accidental Species: Misunderstandings of Human Evolution*. Henry completed his BSc at the University of Leeds and his PhD at Fitzwilliam College, Cambridge.

Jonathan Gibbins is Professor of Cell Biology and director of the Institute for Cardiovascular and Metabolic Research at the University of Reading. He is an expert in the study of the blood cells that trigger the blood to clot following injury. This is important because they also trigger heart attacks and strokes. Jon's laboratory has been responsible for some of the fundamental discoveries that are offering hope in the development of new medicines to prevent and treat cardiovascular disease.

Tim Richardson was awarded a PhD from the University of Reading for work on factors which influence metabolism in the liver. He went on to work at St. Thomas's Hospital, London (cancer research), and the MRC research unit at Harwell (angiogenesis), before joining Amersham International plc to develop various products for life science research. At Amersham he became involved in the management of R&D; in 2004 moved from the commercial sector back into Higher Education and now works in the School of Biological Sciences at the University of Reading.

Tiffany Taylor is a teaching fellow of Biological Sciences at the University of Reading. She is an evolutionary biologist whose research interests include the evolution of gene regulatory networks, the genetic code and the evolutionary ecology of dispersal within the context of cancers (metastasis). She is also an advocate of science communication for both adults and children and has published two children's books on evolution, *Little Changes* and *Great Adaptations*.

Philip J. White obtained his BA from the University of Oxford and his PhD from the University of Manchester. He has published over 300 scientific articles and was included by Thomson Reuters in The World's Most Influential Scientific Minds 2014. He is a member of the International Council on Plant Nutrition, a Full Professor in Biology at King Saud University, and an Honorary Professor at the University of Nottingham. He currently leads a research group at The James Hutton Institute, Dundee, engaged on projects related to plant mineral nutrition and sustainable crop production.

INDEX

ACKNOWLEDGEMENTS

The publisher would like to thank the following individuals and organizations for their kind permission to reproduce the images in this book. Every effort has been made to acknowledge the pictures; however, we apologize if there are any unintentional omissions.

Getty/ Thomas Lohnes: 58.

James King-Holmes/Copyright © James King-Holmes 1996: 46.

Shutterstock/ A7880S: 29TC; Aaltair: 147CT; Abeselom Zerit: 9, 125TC; Adike: 67C, 137C(BG); Africa Studio: 149C; Agsandrew: 15CL&T; ailin1: 147T; Ailisa: 123C(BG); AkeSak: 85C(BG); Alekleks: 141C(BG); Aleksey Stemmer: 127TR; Alesandro14: 49C; alexassault: 103CT&C(BG); Alexilusmedical: 57T,C&B; Alila Medical Media: 37C, 37R; Alslutsky: 77C&B, 129TL; Anan Kaewkhammul: 29TR, 127CR, 137TC; Andrey_Kuzmin: 37C(BG), 129TR, 129CR; Andris Torms: 109C; antoni halim: 127CL; ANURAK PONGPATIMET: 29TL, 127CL; AridOcean: 135CR, 135C; Aromaan: 137C(BG); art4all: 147C(BG); Astronoman: 37C; Attila Jandi: 142; Balein: 19CR, 19TL, 29CL, 57T, 61CL&CR; Bardocz Peter: 135CR; Benny Marty: 119CL; BOONCHUAY PROMJIAM: 129TC; Chad Zuber: 29CL; Charles Brutlag: 81BR; Christian Musat: 135C; Chromatos: 43BC, 43TC, 101C&B(BG); Chuck Wagner: 117CR; Chungking: 151CL; cla78: 117C&BC; Computer Earth: 127CR, 129CR; CreativeNature R.Zwerver: 117CL; D. Kucharski K. Kucharska: 23TR, 31TR; Dangdumrong: 129TL; Daniel Prudek: 151CR; Danny Xu: 25CR, 29TC; Dariusz Majgier: 135C(BG), 145C(BG); David W. Leindecker: 145CL; decade3d - anatomy online: 41C&B; design36: 45C; Deyan Georgiev: 127CR; Dima Sobko: 119BR; Dirk Ercken: 145TR; DK Arts: 27TC, 81C&T; DnD-Production.com: 29TR; Donjiy: 129TR; DrimaFilm: 107C(BG); Dwight Smith: 127TL; Edward Westmacott: 127TC; Ekkapon: 127TL; Elnur: 31CL&CR; Eric Isselee: 89C, 119BC, 119CR, 125CR, 127CR, 129TL, 135CL, 135BR, 147TC; EVo40: 49C(BG); extender_01: 95C; Fedorov Oleksiy: 141B; Filip Fuxa: 17T; Fototehnik: 117BC; GarryKillian: 63C(BG); Gen Epic Solutions: 31C, 89C; Glenn Young: 81BC; Grebcha: 19C&CL, 25C(BG); Haru: 109C(BG); Hedrus: 123C; Hein Nouwens: 65BL, 81BL; Holbox: 127CR; Horiyan: 31C; Horoscope: 61B; HUANSHENG XU: 27CL&CR; Iakov Filimonov: 129TR, 129C; Iamnao: 127TR; Ian 2010: 109CR; Ian Grainger: 149C(BG); Ivosar: 29TC; Jakkrit Orrasri: 29TL; Jbmake: 29TR; Jezper: 85TC, 107TC; Joe White: 23C; Johannes Kornelius: 29TL; jreika: 27C; Juan Gaertner: 63T(BG); Jubal Harshaw: 23C&BL, 27C(BG), 29BC, 81TC; Jukurae: 17C&B; jules2000: 41BL&BR; Juliann: 109BG; jumpingsack: 147T; Justin Black: 145BR; Katarina Christenson: 147T; Kateryna Kon: 19CR, 29TC; Keith Publicover: 79C&B; Khoroshunova Olga: 29C; Kichigin: 15CL; Kletr: 129CL; Kositlimsiri: 145C(BG); Kostyantyn Ivanyshen: 103TL&BR; Kuttelvaserova Stuchelova: 83R&L; Le Do: 27TC; Lebendkulturen. de: 7, 23CL, 23TR, 125C(BG); Leonid Andronov: 95B; LeonP: 125C,
125CL; Lev Kropotov: 27BL; Lightspring: 83C&T, 85C, 103C; Linda Bucklin: 105C; Ljupco Smokovski: 119C; Login: 31TC, 55C(BG); Lukiyanova Natalia / frenta: 55T,C&B; M. Unal Ozmen: 31BL&BR; MAC1: 127CL; Madlen: 99C&T(BG); majeczka: 27TL,TR,&BC; Maks Narodenko: 89C&T, 127CL; Maksym Gorpenyuk: 139C; Marcel Jancovic: 123CR; mariait: 29TR; Markus Gann: 99TC; MARSIL: 147TR; matthi: 151CT; Meister Photos: 135CL; MichaelTaylor: 23TL, 29BC; MichaelTaylor3d: 107CR; microvector: 45C; Mike Truchon: 115TL; Mikhail Kolesnikov: 17C; molekuul.be: 43TL&TC, 43BC&BR, 61C, 95BL&BC, 101C&T(BG); Mopic: 99C, 107C; Morphart Creation: 135CL; motorolka: 139C; Muellek Josef: 129C; Nagel Photography: 79C&T; NattapolStudiO: 117T(BG); Natykach Nataliia: 49C; Nazzu: 141C; Nejron Photo: 29TR, 129CR; NickSorl: 89C(BG); Nixx Photography: 15C, 15CR, 75B&BG; Nowik Sylwia: 49C(BG); O2creationz: 75T,C&B; Ociacia: 31CL&CR; olcha: 129BR; Only Fabrizio: 25C; Onur Gunduz: 19C, 19CL&T, 29CL, 127CL; Oscity: 151C; ostill: 101C; piai: 45C(BG); Pakhnyushchy: 129TC; Pan Xunbin: 23BR, 99C(BG), 127CR; panbazil: 139TC; Pavel L Photo and Video: 151T; PCHT: 117TC, 129TL; petarg: 43C, 95T; PeterVrabel: 119TC; Petr Vaclavek: 101C(BG), 139C; Photo Image: 83C&T(BG); Photobank gallery: 147C; Photoonlife: 127CR, 129TR; Photoraidz: 149C(BG), 151C; Praiwun Thungsarn: 89C(BG); Promotive: 61TR&TC; Protasov AN: 29TC, 79CR; qushe: 67C(BG); RAJ CREATIONZS: 79BC; Ramona Kaulitzki: 65C; rck_953: 147T; Robert L Kothenbeutel: 139TC; Roman Samokhin: 29TR; Rosa Jay: 129TC; S K Chavan: 63C; Sanit Fuangnakhon: 109BC; sciencepics: 23C, 83C; Sebastian Kaulitzki: 41C&B, 63C, 65B, 105CL&CR, 105C&BC, 107CL; shaziafolio: 31CL&CR, 49BL&BR; Slavoljub Pantelic: 109CL; Smit: 29TR; SOMMAI: 83B; stihii:67C; Stock Up: 127TC; stockphoto mania: 25B; Stubblefield Photography: 115CR; subin pumsom: 129CL; Suchatbky: 25C; sutham: 29TL, 127CL; Svetlana Foote: 29CR, 127CR; Talvi: 129CR; Tania Thomson: 129CL; Tatiana Volgutova: 39TL&TR; Tatuasha: 39T,C&B; tdoes: 29CR, 129TR; The Biochemist Artist: 45C(BG); toeytoey: 125T&B(BG); Vaclav Volrab: 79CL; Val_Iva: 29CL; valdis torms: 31C(BG); Valentina Razumova: 139B; Victeah: 31C; Vikpit: 31C(BG), 45C(BG), 49C(BG); Vitoriano Junior: 135CL, 135C; vitstudio: 69C(BG); Vladimir Sazonov: 145TC&BC; Vladislav Gurfinkel: 135C; Voin_Sveta: 147B(BG); Volodymyr Krasyuk: 25C; Vshivkova: 69TC, 69BC, 107C(BG); vvoe: 103C(BG); waniuszka: 104C&B(BG); watchara: 39C(BG), 41C(BG), 43C(BG), 95C(BG), 139C(BG); xbrchx: 123BR; Yure: 55BG; Yuriy Vlasenko: 109CL; Zern Liew: 29C, 127C.

Wellcome Library, London: 19BL, 77T&C(BG).

Wikimedia Commons/Axel Meyer: 41T&C; Cephas: 149CR; Charles H. Smith / U.S. Fish and Wildlife Service: 149BL; James St. John: 149CR; Javier Pedreira: 20; Jim, the Photographer: 149CL; Mateuszica: 127BC; Mike Pennington: 149BR; US Embassy Sweden: 86; Wellcome Images: 37L(BG), 49BG, 65TR&CR, 81T, 137BC.